From the *Foreword*: "This book depicts a 40-year journey beginning when Doug changed his undergraduate major from computer science to mathematical sciences to today where we find Doug as a data science director at Walmart Global Tech, as the co-author of the book *Why Data Science Projects Failv: The Harsh Realties of Implementing Analytics without the Hype*, and teaching practitioners and leaders how to apply analytical science within a business environment. For the duration of his career, Doug has worked at the intersection of mathematics, statistics, computer science, large amounts of data, and real-world problems for both the private sector and the public sector. His journey has had many twists and turns along the way, but the best practices and critical lessons learned that Doug has gleaned from his experiences are invaluable for anyone even tangentially involved with Analytics, Data Science, and Artificial Intelligence today."

Stephen M. Clampett, *Owner of SM Clampett Group LLC, Senior Airline and Travel Technology Executive, Former President Sabre Airline Solutions*

AF736587

The Art of Data Science

Although change is constant in business and analytics, some fundamental principles and lessons learned are truly timeless, extending and surviving beyond the rapid ongoing evolution of tools, techniques, and technologies. Through a series of articles published over the course of his 30+ year career in analytics and technology, Doug Gray shares the most important lessons he has learned – with colleagues and students as well – that have helped to ensure success on his journey as a practitioner, leader, and educator.

The reader witnesses the Analytical Sciences profession through the mind's eye of a practitioner who has operated at the forefront of analytically inclined organizations, such as American Airlines and Walmart, delivering solutions that generate hundreds of millions of dollars annually in business value, and an educator teaching students and conducting research at a leading university. Through real-world project case studies, first-hand stories, and practical examples, we learn the foundational truth underlying successful analytics applications. From bridging theory and practice, to playing a role as a consultant in digital transformation, to understanding how analytics can be economically transformational, identifying required soft skills like leadership skills, and understanding the reasons why data science projects often fail, the reader can better visualize and understand the nuanced, multidimensional nature of Analytical Sciences best practices, projects, and initiatives.

The readers will gain a broad perspective on where and how to find success with Analytical Sciences, including the ability to ensure that we apply the right tool, at the right time and right place, and sometimes in different industries.

Finally, through the author's own career synopsis on becoming a practitioner and leader, and his distilled insights, the reader is offered a view into the future that analytics holds, along with some invaluable career advice regarding where to focus, how to make good choices, and how to measure success individually and organizationally.

The Art of Data Science

A Practitioner's Guide

Douglas A. Gray

CRC Press
Taylor & Francis Group
Boca Raton London New York

CRC Press is an imprint of the
Taylor & Francis Group, an **informa** business
A CHAPMAN & HALL BOOK

Designed cover image: Getty Images_1367899893

MATLAB® and Simulink® are trademarks of The MathWorks, Inc. and are used with permission. The MathWorks does not warrant the accuracy of the text or exercises in this book. This book's use or discussion of MATLAB® or Simulink® software or related products does not constitute endorsement or sponsorship by The MathWorks of a particular pedagogical approach or particular use of the MATLAB® and Simulink® software.

First edition published 2025
by CRC Press
2385 NW Executive Center Drive, Suite 320, Boca Raton FL 33431

and by CRC Press
4 Park Square, Milton Park, Abingdon, Oxon, OX14 4RN

CRC Press is an imprint of Taylor & Francis Group, LLC

© 2025 Douglas A. Gray

Reasonable efforts have been made to publish reliable data and information, but the author and publisher cannot assume responsibility for the validity of all materials or the consequences of their use. The authors and publishers have attempted to trace the copyright holders of all material reproduced in this publication and apologize to copyright holders if permission to publish in this form has not been obtained. If any copyright material has not been acknowledged please write and let us know so we may rectify in any future reprint.

Except as permitted under U.S. Copyright Law, no part of this book may be reprinted, reproduced, transmitted, or utilized in any form by any electronic, mechanical, or other means, now known or hereafter invented, including photocopying, microfilming, and recording, or in any information storage or retrieval system, without written permission from the publishers.

For permission to photocopy or use material electronically from this work, access www.copyright.com or contact the Copyright Clearance Center, Inc. (CCC), 222 Rosewood Drive, Danvers, MA 01923, 978-750-8400. For works that are not available on CCC please contact mpkbookspermissions@tandf.co.uk

Trademark notice: Product or corporate names may be trademarks or registered trademarks and are used only for identification and explanation without intent to infringe.

Library of Congress Cataloging-in-Publication Data
Names: Gray, Doug (Douglas A.), author.
Title: The art of data science : a practitioner's guide / Douglas A. Gray, MSOR, MBA.
Description: First edition. | Boca Raton, FL : CRC Press, 2025. |
Includes bibliographical references and index. |
Identifiers: LCCN 2024041228 | ISBN 9781032818177 (hardback) |
ISBN 9781032816968 (paperback) | ISBN 9781003588344 (ebook)
Subjects: LCSH: Business analysts. | Management–Statistical methods. | Quantitative research.
Classification: LCC HD69.B87 G73 2025 | DDC 658.4/012–dc23/eng/20241107
LC record available at https://lccn.loc.gov/2024041228

ISBN: 9781032818177 (hbk)
ISBN: 9781032816968 (pbk)
ISBN: 9781003588344 (ebk)

DOI: 10.1201/9781003588344

Typeset in Palatino
by codeMantra

To my parents, my wife, and my sons, without whose love

and support I would have achieved far less.

To all of my professors at Loyola University, Georgia Tech,

and Southern Methodist University who provided me with

the educational foundation for my professional life.

To all of my leaders, colleagues, and students throughout my career

with whom I have been on the journey of delivering business value and

economic impact and continually learning through failure and success.

To the current and future generations of data science, analytics,

AI, operations research (O.R.), and statistics practitioners

and leaders for whom this book is intended.

Contents

Foreword

With analytics, data science, and artificial intelligence (ADSAI, as the author puts it) fast becoming key components of corporate strategy and most technology projects, this book is a must read for anyone who is engaged either directly or indirectly with work or research in these disciplines or for anyone who is just curious about what all the hype is about around ADSAI. Through the lens of his own career and through the different papers and articles he has published through the years, Doug Gray does a masterful job of creating a mosaic of how applied mathematics, statistics, operations research, and management science have evolved into the disciplines of analytics, data science, and artificial intelligence that we know today. This mosaic depicts a 40-year journey beginning when Doug changed his undergraduate major from computer science to mathematical sciences to today where we find Doug as a data science director at Walmart Global Tech, as the co-author of the book *Why Data Science Projects Fail: The Harsh Realties of Implementing Analytics without the Hype,* and teaching practitioners and leaders at the Southern Methodist University Cox School of Business how to apply analytical science within a business environment. For the duration of his career, Doug has worked at the intersection of mathematics, statistics, computer science, large amounts of data, and real-world problems for both the private sector and the public sector. His journey has had many twists and turns along the way, but the best practices and critical lessons learned that Doug has gleaned from his experiences are invaluable for anyone even tangentially involved with ADSAI today.

Doug purposely makes the audience for this book very broad. For the high school or college student contemplating pursuing a degree in one of the fields of ADSAI, Doug shares in detail his own decision process for pursuing an undergraduate degree in mathematical sciences at Loyola University Maryland and a graduate degree at Georgia Tech. The self-assessment he put himself through to decide that the field of operations research was best for him and the way he proactively reached out to his professors for advice and counsel are best practices for any student pursuing a degree or career in ADSAI. In fact, I would argue that these are best practices for any student choosing a major or career path no matter what the area.

For the ADSAI practitioner, there are key lessons learned and best practices sprinkled throughout this book. Doug devotes a whole chapter in the book to the non-technical skills (soft skills) that make an ADSAI practitioner successful. Anyone contemplating a career in this area should take a hard look in the mirror and ask themselves if they are interested and motivated to work hard and to develop their skills in these requisite non-technical areas. As Doug walks the reader through his own career progression, he provides wonderful nuggets of career advice. One such counterintuitive nugget is found in Chapter 10 where Doug says, "My MBA made me a better data science practitioner." This particular point struck a nerve with me personally. I graduated from the University of Missouri-Columbia with an MS in Applied Mathematics and then went to the Purdue University Daniels School of Business where I received an MS Industrial Administration degree, which I called at the time an "MBA for nerds." For the same reasons Doug articulates in Chapter 10, I always say my business degree from Purdue was one of the best career decisions I ever made. At a higher level, Doug's willingness to do the dirty work and take on assignments that others shied away from, his obsession with thoroughly understanding the business problem he is trying to solve, and the deep relationships he developed with organizational sponsors and mentors are not only great career advice for the ADSAI practitioner but also great advice for people across many different careers.

For the leaders of large or small ADSAI organizations, Doug dedicates Chapter 11 to you. This chapter is based on a paper Doug co-authored with Tom Davenport, and the over-arching theme of Chapter 11 is that the skills that got you to a leadership role are not the skills that will make you successful in that role, which is something that many people with a strong technical background have a difficult time wrapping their head around. In the paper, the authors list 10 key attributes of a successful ADSAI leader. While all 10 are very important, the first two on the list are critical. Topping the list is recruitment, retention, and people development. Since people are the only raw material of an ADSAI organization, how you recruit, retain, and develop these people is obviously fundamental to the success of any ADSAI organization. The second critical skill on the list, generating demand (securing projects by domain area), is not quite so obvious. To highlight the importance of this skill, Doug tells a story about how he had a boss (GM Jeff Honeycomb at McAfee) who would say, "Nothing happens until somebody sells something." Clearly, there is no work for an ADSAI organization to do until a sponsor in a business unit or an outside client asks for the work. This skill is also one that many technical folks loathe to develop and is why many great technicians fail when put into leadership roles. Any person moving into a leadership role for the first time will not be adept in all 10 skills articulated in Chapter 11. The question a person needs to ask when contemplating a move into a leadership position is: Do I have the intellectual curiosity and personal desire to develop and grow across these 10 dimensions as a leader? If the answer is "No," Doug's vast experience chronicled in this book would say that a person is better off staying on a technical career path.

For the business unit leader/sponsor for an ADSAI project or initiative, Doug is very clear throughout the book that the success of the project hinges on its ability to drive significant and measurable business value, which cannot happen without the active participation of your team throughout the entire life cycle of the project. Doug makes a compelling case that at the heart of any successful ADSAI project is a tight partnership between the business unit and the technology teams. As a business unit sponsor, you should take note of the top 10 reasons projects fail in Chapter 12 and do everything you can within your organization to minimize these risks.

Doug also spends a significant amount of time in the book talking about digital transformation. If you "google" digital transformation, you get a myriad of results from different consulting firms and technology companies. Clearly, digital transformation is a hot area these days. As I read Doug's career journey in this book, what struck me is that he has been part of a digital transformation for different organizations for the past 40 years. While different components of a digital transformation can be broken into projects and managed like projects, the transformation itself is a journey much like Doug's career with many twists and turns along the way. As a leader of such a journey, the best piece of advice for you can be found at the beginning of Chapter 12 where Doug highlights Stephen Covey's "The 7 Habits of Highly Effective People" and focuses on "Begin with the end in mind." There is no better advice for a leader embarking on a digital transformation than these simple six words!

For the researcher in one of the fields of ADSAI, Doug's journey highlights how massive improvements in compute power and data storage have provided us the ability to solve problems today that we only dreamed about 30–40 years ago. As a prime example, Doug highlights the airline real-time irregular operations problem that he and a small team addressed at two different points in his career. In the early 1990s, Doug and a team tried and quickly failed to solve this problem for American Airlines. Twenty to twenty-five years later, Doug and his team at Southwest Airlines solved this problem driving significant, measurable business value that Southwest Airlines still realizes today. Through the many examples Doug references, he makes the case that in most instances, a robust technical solution is a necessary condition for a successful ADSAI project. He also makes the case, however, that while these technical solutions can be quite sophisticated, they are not a sufficient condition for a successful project. Consequently, it is incumbent upon the researcher that pushes the state-of-the-art across different technical boundaries to be cognizant of what ultimately makes a successful ADSAI project and understand how the research fits or does not fit within that framework.

Finally, for a grizzled veteran like myself, who has spent the last 45 years of my life in many different roles applying operations research techniques and developing decision support systems, *The Art of Data Science: A Practitioner's Guide* provides me an opportunity to look back on my own career, to take pride in what we accomplished, to think about the many mistakes we made and what we learned from those mistakes, and most importantly to reflect on where the disciplines of analytics, data science, and artificial intelligence go from here. My favorite chapter in the book is Chapter 14 entitled "O. R. in 2048." The crux of the chapter is an article published in 1998 in *OR/MS Today* where Peter Horner, the editor of the magazine, asked Doug and a few other industry professionals to provide their perspective on what they thought O. R. (Operations Research) would be 50 years in the future, in 2048. Reading the article again in 2024, 26 years later, I must say Doug was directionally right across many dimensions, but probably a little too conservative in some areas. If you take stock of where the fields of analytics, data science, and artificial intelligence are today versus where industry professionals thought they would be 26 years ago, it bodes well for where they will be, or where their derivative disciplines will be, in 2048. I hope you enjoy this chapter as much as I did!

In this book, Doug describes eloquently how his formal education from Loyola University Maryland, Georgia Tech, and SMU Cox Business School and the subsequent undergraduate and graduate degrees he received from these institutions provided him with a very strong foundation for his career. But he goes on to say, "My doctorate is from the school of hard knocks." It is from this "school of hard knocks" where the invaluable best practices and lessons learned emerge as Doug walks us through his career. I have known Doug since he joined American Airlines Decision Technologies (AADT) in 1987, and I have a tremendous amount of respect for his capabilities and what he has accomplished in his career. I cannot think of a better person to write a book like *The Art of Data Science: A Practitioner's Guide*. Enjoy Doug's journey, and please take the time to digest the many best practices and critical lessons learned that Doug provides you along the way. Finally, and most importantly, stop and reflect on how you can best use these key learnings in your own situation.

Stephen M. Clampett

Stephen (Steve) Clampett is an internationally recognized airline and travel technology executive with a proven track record of building successful businesses.

Steve is the former President, Airline Products and Solutions for Sabre Holdings, Inc. In this role, he had P&L responsibility for Sabre Airline Solutions, a $500+ million Sabre business unit which had the strongest portfolio of decision support systems in the airline industry and a customer base of 200+ airlines and airports worldwide. His responsibilities included marketing, developing, and delivering solutions in areas such as reservations, flight scheduling, crew management, flight operations, pricing, revenue management, cargo, and revenue accounting for Airline Solutions' clients worldwide. Steve was instrumental in the transformation of Sabre Airline Solutions from a custom development organization based in Texas into a global product-focused business with offices in the United States, India, Poland, and Uruguay. Under Steve's leadership, Sabre Airline Solutions launched a Software as a Service (SaaS) delivery model for airline decision support products, which was instrumental in transforming the business into a recurring revenue model.

After beginning his career with the Operations Research Department at Ford Motor Co., Steve joined American Airlines as an operations research analyst and later became Vice President of American Airlines Decision Technologies (AADT), a wholly-owned subsidiary of AMR Corporation. AADT later grew and evolved into Sabre Airline Solutions.

Steve currently is the owner of SM Clampett Group LLC, an airline and travel distribution consulting firm, and an operating advisor for Liberty Hall Capital Partners. Previously, he was a Board Advisor for Comply365 and a member of the Board of Directors (Audit and Compensation Committees) for GuestLogix Inc.

Steve holds a master's degree in Industrial Administration from the Mitch Daniels School of Business, Purdue University and master's and bachelor's degrees in Applied Mathematics from the University of Missouri. Steve was also a member of the University of Missouri College of Arts and Sciences Strategic Development Board from January 2009 to July 2021 and the University of Texas at Dallas School of Management, Supply Chain Management Industry Advisory Board from June 2011 to June 2023.

Acknowledgments

To my publisher at Taylor & Francis, Ms. Randi Slack, whose support, encouragement, guidance, and direction on both of my books have been unwavering from the very first conversation and so very instrumental in completing these writing projects. Randi took a chance on a first-time author and looked past my inexperience to focus on the content of the message of someone who believed they had something important to say. Her judgment, insight, publishing business acumen, and magical words, *"We have to publish this book!" (twice)* have made all of the difference. (And she was instrumental in the titling of both books!)

To my editor, Ms. Kara Tucker, for inviting me to publish, and then editing, my INFORMS *Analytics* magazine 12-part article series on *The Top 10 Reasons Why Data Science Projects Fail,* which then gave rise to the book *Why Data Science Projects Fail* (with my co-author Evan Shellshear) and *The Art of Data Science,* both of which she also edited. Thankfully, because I am the furthest thing possible from a literary grammarian or book production specialist.

To my co-author on the first book, Evan Shellshear, thank you for helping to make me a better writer and a more rigorous editor.

To my first software engineering partner, Bobby Johns, on my very first big, successful project at AA, highlighted in Chapter 5, that materially helped to launch *both* of our careers, a debt of gratitude for making my O.R. model more easily consumable and intelligible to end users with super cool GUIs, and for making my code run faster and more efficiently. (Bobby was also my chief engineer on *Travelocity,* which he helped to make into a *huge* success, and multiple other projects with me over the years, as well as a great friend who always made work more fun!)

To Dr. Phil Beck, the manager of my Optimization Solutions group at Southwest Airlines where together we garnered multiple industry awards and 9-figures of validated business value annually delivering game-changing O.R., analytics, and data science solutions across the airline operations spectrum. Phil is a true colleague and friend whose partnership I value greatly.

To Nader Kabbani, my co-author on *Right Tool, Right Place, Right Time* in Chapter 8, thank you for your partnership.

To Dr. Peter C. Bell, Professor Emeritus, Management Science, at Western University's Ivey School of Business, for permission to quote and reprint his *OR/MS Today* article "Defining Analytics Through the Eyes of Students" in Chapter 13. Dr. Bell taught a course called Competing with Analytics to EMBA students (which is very similar to the course I teach at SMU!). He is a past recipient of the INFORMS Prize for the Teaching of OR/MS Practice.

To all of my colleagues throughout my career, whose part in these stories are duly noted and included in this book. To my world-class teams and team members at American Airlines, Southwest Airlines, and Walmart Global Tech – there are too many to name here, but you know who you are. I couldn't have done it without you. Thank you!

About the Author

Douglas A. Gray is a practitioner, leader, educator, author, and advisor with 30+ years of experience in data and analytics, IT/software, e-commerce, and consulting. Doug is the co-author of *Why Data Science Projects Fail: The Harsh Realities of Implementing AI and Analytics, without the Hype* with Dr. Evan Shellshear (Taylor & Francis – CRC Press). He and his teams have won several industry awards, including the Teradata EPIC Award, FICO Decision Management Innovation Award, Alteryx Best Business ROI Award (2), Drexel LeBow Analytics 50 Award (2), and AGIFORS Operations Best Innovation Award. Doug and his teams annually deliver 9-figures validated business value and economic impact to the enterprise and have cumulatively delivered more than $3 billion throughout his career.

He has been a member of two INFORMS prize-winning organizations: Walmart (2023) and American Airlines (1990). Doug is currently a Director, Data Science at Walmart Global Tech – US Omni Retail Tech, where he leads teams and programs that increase operational efficiency and cost effectiveness focused on end-to-end fulfillment. Doug was an early team member at award-winning American Airlines Decision Technologies (AADT), which pioneered numerous innovations driving substantial economic impact in the use of analytics in the airline industry, solving commercial and operational problems across the enterprise. At Sabre, Doug served as the founding CTO of Travelocity, the world's first real-time, internet-based travel reservation system.

Through his own company, Blueprint Technology Advisors, LLC, d.b.a., Optima Analytics, he teaches and advises executive leadership teams on best practices of applying technology, data, and analytics to corporate strategy, operations, and digital transformation and implementing analytics organizations and capabilities.

Doug has been teaching Business Analytics and AI Strategy at SMU's Cox School of Business and SMU's MS in Data Science program as an adjunct professor since 2016. He has published over a dozen articles and research papers on analytics applications and has been an invited keynote, guest speaker, and panel discussion participant at industry conferences and universities worldwide.

Doug holds an MBA from SMU's Cox School of Business (Beta Gamma Sigma), MS in Operations Research from the Georgia Institute of Technology's Stewart School of Industrial & Systems Engineering, and BS (cum laude) in Mathematical Sciences (Statistics) from Loyola University Maryland (Pi Mu Epsilon). Doug is a recognized Analytics Expert at the International Institute for Analytics (IIA).

Introduction

Motivation

This book is an important part of what I call my professional "Legacy Project," which is founded on how I want to spend the final phase of my formal work life while transitioning into retirement, and what I want to leave behind for others to learn from my decades-long career.

My professional Legacy Project consists of the following components:

- Writing a book, or two, that shares important timeless lessons learned on the practice of the analytical sciences, i.e., data science, analytics, Artificial Intelligence (AI), operations research, and statistics.
- Working for a world-class company known for aggressively applying data science as a part of digital transformation, delivering value, and leading and mentoring other practitioners (i.e., Walmart Global Tech).
- Teaching practitioners and leaders in a top university setting how to apply analytical sciences within a corporation (i.e., Southern Methodist University (SMU) Cox School of Business EMBA and Executive Education Programs, Continuing & Professional Education (CAPE) Programs, including the MS in Data Science).
- Consulting and advising companies on how to more effectively apply analytical sciences to problem-solving, decision-making, and question answering (i.e., my own company, Blueprint Technology Advisors, LLC, d.b.a., Optima Analytics).
- Sharing what I have learned through speaking engagements at a wide range of venues, large and small, e.g., conferences, symposia, universities, and leadership and practitioner gatherings.

I was inspired to write books about the practical application of analytics, AI, and data science by Tom Davenport, Ph.D., whose prodigious and prolific authoring of books on the subject, including the seminal works *Competing on Analytics* and *All -iIn on AI*, helped to immeasurably shape my own thinking

DOI: 10.1201/9781003588344-1

on my professional field of endeavor unlike any other author. (I have read *all* of Tom's books on analytics and AI. I first spoke at Tom's International Institute for Analytics Conference in 2017 on the work I was leading at Southwest Airlines, and we have since collaborated and kept in touch, and I am grateful for his friendship and for writing the Foreword of my first coauthored book, *Why Data Science Projects Fail.*)

My first book, *Why Data Science Projects Fail: The Harsh Realities of Implementing AI and Analytics, without the Hype*, coauthored with Evan Shellshear, and publisher Randi Slack at Taylor & Francis, is *foundational* to my Legacy Project because it is unique and distinct in that it focuses on the *failures of others* in their endeavors, which is how people learn most effectively. (There is no "mental block" to surmount and accept that you failed.) In 2022–2023, I published a 12-part series of articles on "The Top 10 Reasons Why Data Science Projects Fail" in INFORMS *Analytics* magazine, with editor Kara Tucker. The series provided the initial impetus for my first book. In addition to our own failure stories, Evan and I researched many, many others and documented why those projects failed, what lessons could be learned, and what could be done differently to help others ensure success more frequently and more consistently.

During the publication of *Why Data Science Projects Fail*, I was assembling a CV, which included a list of all of my publications, public speaking, courses taught, etc. In the 1990s, I published a half-dozen articles in *OR/MS Today* (i.e., the flagship magazine of INFORMS – The Institute For Operations Research and Management Science), with then-editor Peter Horner, all about analytics (at the time, still referred to as operations research) practices, principles, and projects. In writing my CV, I reread all of those articles and realized there were a lot of timeless lessons that would still be valuable to practitioners and leaders in today's world.

More recently, I published another half-dozen posts on LinkedIn regarding leadership skills and the transformational nature of analytics that many people liked and commented on as being insightful and helpful. Upon review, Randi agreed that there was indeed a worthwhile story in the collection of articles, posts, and some additional insights and rubrics to share with the next generation of practitioners and leaders.

An extra important bit of context for younger readers or those new to the analytical sciences – operations research (O.R.), which originated with military operations analysts during WWII in the 1940s and advanced in the 1950s in academia and business, is a discipline that deals with the development and application of analytical methods to improve problem-solving and decision-making. O.R. (and its synonymous and practically contemporaneously founded the discipline of management science, i.e., the business use of O.R.) has undergone a branding and marketing transformation in recent years to be closely associated with the field of analytics and is application- and mission-wise adjacent to and closely aligned with the fields of data science and artificial intelligence (AI) (including machine learning, of course), which evolved from statistics and computer science, respectively. For more detailed information on O.R., its origin, and evolution and its myriad problems addressed, modeling methods, and applications, see: https://en.wikipedia.org/wiki/Operations_research.

I am proud to say that my career is "bookended" by working in O.R. and data science, respectively, for two INFORMS Prize- and INFORMS Franz Edelman Award-winning organizations: American Airlines (Decision Technologies) and Walmart (Global Tech). (To be clear, I was not a member of *either* Edelman Award-winning team, but the awards speak to the caliber of the cohorts of which I was a member.) At those companies, and others in between, including Sabre, Blue Cross, and Blue Shield of Kansas City (on contract as Acting Director of Analytics), and Southwest Airlines, I personally did a lot of good work as a practitioner and led teams that did a lot of good work with analytics that led to well over *$3 billion* (at the time of publication) in cumulative, Finance department-validated business value and economic impact. This book will address in detail portions of that value that can be made public (as most of it cannot). That value and impact is an important component of my *legacy*, because it is a testament to the cadre of people that I recruited, hired, trained, taught, and learned from myself while delivering projects.

Creating a Course and a Legacy

In mid-2014, unsolicited, I developed a proposal for a 1.5 credit Business Analytics elective course for Executive MBAs at SMU's Cox School of Business, of which I am also an alumnus (EMBA '06), to teach executives how best to incorporate analytics into their corporate strategy and tactical problem-solving, decision-making, and question answering. Despite having *no prior formal teaching experience at any level*, Tom Perkowski (EMBA Program Assistant Dean) gave me an opportunity to teach at SMU Cox. Starting in 2016, I taught that course five times to about 120 students in total. That experience is an important part of my *legacy*, in that many of those current and future executives in many instances *changed the trajectory* of the companies for which they worked, and identified, in total, hundreds of millions of dollars of business value and economic impact by applying analytics using the methodologies and rubrics I taught them.

In early 2019, SMU's MS in Data Science Program Director, Dr. Bivin Sadler, asked me to create a 3-credit Business Analytics elective course for their students, based on the course I developed for the business school. We recorded my lectures and that course is taken every year by a majority of the cohort in MS in Data Science program. That experience is an important part of my *legacy* in that many of those current and future practitioners, managers, and executives in many instances will be better equipped to apply data science in their respective enterprises.

In 2025, I will be offering a new elective course that I proposed and created, again unsolicited, on AI Strategy in SMU Cox's EMBA Program to help executives understand what AI *really is*, beyond the 24-hour news cycle hype machine about ChatGPT and robots taking over the world, and how it can be practically applied to solve complex problems, make better decisions, and answer questions more holistically in the enterprise. I am looking forward to being back in the classroom after my sabbatical from teaching to write two books. In January 2024, I began teaching a one-day seminar in SMU's Cox Leadership Academy (Executive Education) on Analytics & AI Strategy, leveraging the material developed for my two courses just described, as well as delivering seminars and lectures on AI to a range of business and technical audiences; the subject of AI continues to evolve as the new hot topic.

In 2009, while living in Scottsdale, Arizona, in the wake of the 2008 financial crisis and economic meltdown, I started my own LLC offering Chief Technology Officer and analytics advisory and consulting services. I made a living and supported my family that way for about five years until I returned to Dallas to lead Optimization Solutions, and then Enterprise Data & Analytics, at Southwest Airlines. Since then, I have continued to operate my LLC and provide analytics consulting, advisory, and executive education teaching services "part-time on the side" from my "full-time day job." When I retire from corporate America, I will continue to operate my LLC, part-time, and provide analytics consulting, advisory, and executive education teaching services, and teach at SMU, if they need me, because teaching and advising others to transform their enterprises and organizations with analytics and data science, to create tremendous business value and economic impact, and creating strategic competitive advantage, is my *passion*. Sharing all that I have been so fortunate to learn, achieve, and experience, through my own successes and failures, with others who embody that same desire is an important part of solidifying my *legacy*.

Organization

The Art of Data Science: A Practitioner's Guide is intended for current and future students, practitioners, and leaders studying and working in the analytical sciences. The goal is to provide insight into *best practices* for applying analytical sciences as well as career progression and development, and how to maximize the business value and economic impact that data science, analytics, O.R., statistics, and AI can offer enterprises. These best practices stem from my own experiences executing projects and leading teams, and those of my colleagues and students, over a 30-year career. The "hook" in the title is the word *"art."* The juxtaposition of "art" versus "science" is intended to focus the reader on so many of the nontechnical *"soft skills"* that make *all of the difference* in successful applications of the analytical sciences. This is not a technical textbook, rather a book addressing some of my methodologies, approaches, and rubrics that are based on experience and "know-how" versus sophisticated mathematical theory.

The book is organized as follows:

- Chapter 1 is a summary of my career progression from a practitioner to leader to educator, the projects on which I worked, the experiences I had, some of the people who played a key role in my development, and most importantly, some of the important lessons that I learned along the way that shaped my professional development.
- Chapter 2 covers some of the early principles that I discovered in surveying the opportunities for practitioners that formed a foundation for all of my future work in the field.
- Chapters 3 addresses digital transformation and the important role analytical sciences play in that domain, which is so ever-present and critically necessary in the economy and corporations in the 21st century to become more economically efficient and provide better customer experiences.
- Chapter 4 delves into how analytics, data science, and AI (ADSAI) as a domain is economically transformational and provides some examples across industries and companies of the metrics affected by analytics.

The *arc* of Chapters 5–9 represents some of the most impactful modeling and solution approaches and projects on which I worked throughout my career, and how those experiences led to foundational principles and lessons learned that I applied with success time and time again:

- Chapter 5 provides a detailed summary of the first model/system solution that I developed from "scratch," along with a super-talented software architect/engineer (Bobby Johns, at American Airlines), the methodologies and technologies we utilized, the human dimension of the project and solution, and the business value and economic impact that we generated.
- Chapter 6 covers many of the lessons learned on the project in Chapter 5, and how those lessons became foundational and repeatable templates for future models, solutions, projects, and success.
- Chapter 7 provides a powerful example of how analytics can be economically transformational and demonstrates how the lessons learned and principles imbued in the project described in Chapter 5 were directly applicable to a different business problem.

- Chapter 8 relates how my coauthor Nader Kabbani (later a highly successful VP at Amazon) and I approached the domain of planning, scheduling, and operations problems and their analytical solution approaches using the airline context as the "lens" to "envision" the characteristics of these problems, and how best to address them.
- Chapter 9 addresses my attempt to apply airline revenue/yield management concepts, principles, methods, technologies, and solution frameworks to a completely different industry, i.e., manufacturing, and brings home the undeniable fact that sometimes an idea is a bit too early and arrives before its time, but then winds up being implemented 25 years later by someone else who you taught and influenced, not to mention *countless others,* as the manufacturing product mix/production capacity allocation optimization problem.

The *arc* of Chapters 10–12 offers a collection of skills and key learnings that I have developed throughout my career and found invaluable as a practitioner, leader, and educator:

- Chapter 10 addresses the nontechnical "soft skills" that are *critical* to develop and complement the technical skills learned in undergraduate and graduate schools.
- Chapter 11 addresses 10 fundamental leadership skills for those seeking to manage and lead others and operate successful ADSAI departmental functions.
- Chapter 12 addresses my top 10 reasons why data science projects fail and was the seed of an idea that led to my first book, coauthored with Evan Shellshear (*Why Data Science Projects Fail*).
- Chapters 13 and 14 highlight where and how to find success and look toward a view of what the future of O.R., and more generally ADSAI, holds:
- Chapter 13 complements Chapter 12 with a holistic view of where to find success in broad-based terms with ADSAI.
- Chapter 14 is my attempt at futurism from back in 1998, in the form of a 50-year forecast projecting what O.R. would be in *2048;* now that we are halfway there, let's see how accurate my forecast was, or wasn't, and look at what's changed since 1998!

The Conclusion sums up the overarching messages and underlying principles of the *art of data science* and hopefully leaves the reader better prepared for their career opportunities and challenges in the analytical sciences.

1

Career Summary: On Becoming a Practitioner and Leader

As a practitioner and leader, I am – professionally speaking – probably a lot like you.

When I was a kid in school, I loved math and problem-solving, and I was pretty good at both – math came easily and naturally to me. In my free time, I enjoyed doing the logic puzzles where you are given several seemingly unrelated statements and then asked a question that you could answer by utilizing deductive reasoning and organizing the information into a matrix.

In junior high and high school, I took the academic math-track courses (i.e., Algebra I and II, Geometry, Trigonometry, Analytic Geometry, College Algebra, and Calculus I and II) and was *blessed* with a group of wonderful math teachers who really took an interest in their students and had a passion for the material. In 11th grade, I struggled a bit with College Algebra because it was the first time in my life that I had to actually study to get good grades – the material was quite abstract (e.g., functions, groups, fields, and rings) and more complex than anything I had seen before. (I was also playing football that semester and chasing my girlfriend around Baltimore when I should have been cracking open my books!)

In my high school "Computer Math" class (1981–1982), I discovered my love for computer programming. We wrote programs calculating, for example, descriptive statistics on data sets on an IBM mainframe computer that a company had donated to Loch Raven High School in northeastern Baltimore County. (The computer took input using the Beginners' All-purpose Symbolic Instruction Code (BASIC) language from "bubble" paper punch cards that we marked up using a Sharpie. We took a school bus there once per week to run and debug our programs.) We also had access to an Apple IIe PC in the classroom on which we wrote some of our assigned programs. We were required to do a final class project, and I chose to write a program for playing the card game Blackjack (21) using subscripted variables and a random number generator to "shuffle" the deck. Programming taught me to pay attention to detail and that computers don't make mistakes – programmers do!

There was something exciting, to me at least, about the sensation of running your computer code, seeing whether it worked and delivered the end result, and then, when necessary, debugging the code to get it to work. Debugging was like a puzzle, a *mystery* to be solved, and further developed

DOI: 10.1201/9781003588344-2

my interest in complex problem-solving, but with a powerful digital tool at my disposal. My classmates and I had friendly competitions on each assignment to see who could get their code working with the fewest number of debugging iterations. Those experiences taught me to spend more time on program design *prior* to writing and running the code – a lesson that served me very well throughout my professional career in building models and systems.

About the time I graduated from high school in 1982, computers and computer science (CS) were just beginning to take off. Many people were opting to major in CS in college. I had a friend who was attending Virginia Tech majoring in CS and was making "good money" working for the U.S. Federal Government at Aberdeen Proving Grounds in Maryland. I decided to take that route, so when I got to Loyola College (now Loyola University Maryland), I majored in CS.

Well, I got a rude awakening. "Introduction to Computer Science" was the freshman CS major "weed-out" course. The professor – a cranky physics professor turned programmer extraordinaire who had a reputation for despising freshmen – announced on Day 1 to a class full of 50 wannabe CS majors that his job was to "get rid of half of you by the end of the semester." He was successful – half of the class failed and were told to "find a new major," most of whom made haste for the business school Management Information Systems (MIS) major.

Fortunately, I passed the class (barely), but I was not feeling super confident in my programming skills relative to the newfound competition. So, I continued taking my CS major courses (Data Structures, Programming Languages, and Physics) but not with the gusto I arrived with. I started to look at the upper-level CS courses in the catalog (e.g., Compiler Construction, Operating Systems, Database Systems, and Digital Design). These topics did not excite me *at all*, and in my sophomore year, I did some deep soul-searching and considered changing my major – but to what? I didn't want to follow the herd to the business school to get an MIS degree.

I did a self-assessment, talked to my advisor (a wonderful Jesuit priest, Father John Brunett, who asked me a lot of great questions), and took inventory of what I was *really interested* in doing.

- I liked computer programming, but not enough to write compilers or operating systems.
- I loved solving problems using mathematics and statistics.
- I was intrigued by the notion of businesses – what made them tick, how they make a profit by serving customers, etc.
- I liked working with people on group projects.
- I absolutely *loathed* the idea of sitting in front of a computer writing code all day.

The "Key" to My Future

I had the great fortune of having an *awesome* professor for Calculus I, who was also chairman of the Mathematics Department at the time (Dr. John C. Hennessey). Originally from New York City (B.S. from Fordham University, M.S. from Purdue University, and Ph.D. from University of Maryland, College Park), Dr. Hennessey was "old school," tough, and very demanding (e.g., he failed half of my Calculus I class; I thankfully emerged with a solid B). He was also very passionate about the application of mathematics to real-world problem-solving. He worked part-time as a mathematician-statistician at the Social Security Administration in Woodlawn, Maryland, working on projects like predicting whether, and when, someone injured on the job and temporarily collecting full disability payments would return to work (i.e., a binary classifier using logistic regression). Sound familiar? Such models had significant implications on planning required cash reserves for future disability claim payouts.

When I told him that I was thinking of changing my major, but was unsure of what I wanted to do career-wise, and shared my interests, he spun around in his chair turning to his massive bookshelf – that seemed to be creaking under the load of hundreds of heavy math books – and pulled out and handed me a thick black textbook with the words **Operations Research** stenciled down the spine. The book was by Fred Hillier and Gerald Lieberman, the seminal introductory textbook on the field of operations research, and used at the time to teach Operations Research I and II at Loyola, among myriad other institutions. Dr. Hennessey said, "Take this book home and read through it over the weekend and see if you are interested in this subject. I seriously believe that this would be a great fit with your interests and skills."

Throughout a person's lifetime, especially early on, there are a few pivotal events that dramatically alter the path one will take and the trajectory of the arc one will follow. Poring through the *Operations Research* textbook that weekend was such a transformative experience for me. Reading about all of the applications of mathematical models and algorithms, e.g., linear programming and queueing theory, to solve practical real-world problems in manufacturing, transportation, healthcare, etc., instilled with a level of excitement and enthusiasm about my education and career opportunities, prior to which I had not experienced. It was as if Dr. Hennessey had handed me a "key" to unlock my future, and indeed, it did.

When I returned the book on Monday and told him what I experienced when reading through the material, Dr. Hennessey fished through his desk drawer and handed me two sheets of paper: One was a "Change of Major" declaration form to switch my major to mathematical sciences, which I did immediately, and the second one was a flyer for an event called "Career Night in the Mathematical Sciences" (which was basically free beer and informal presentations by professionals who worked in businesses that use math – hard to pass up for a college student who had just discovered a new career path).

When I showed up for career night, I didn't really know what to expect. I snuck into what was essentially an oversized conference room (i.e., Cohn Hall under the Loyola Alumni Chapel) with a bit of timidity as someone who was not quite sure that they really belong here just yet. I grabbed a beer and found a place to sit in the back row. What I heard over the next few hours convinced me that I was indeed in the right place. The speakers, many of whom were graduates of Loyola's Math Department, ranged from actuaries who worked at insurance companies to manufacturing plant engineers who worked for steel companies or automobile companies and to others who worked for Department of Defense (DoD) contractors (i.e., "Beltway Bandits"), which were quite common in the Baltimore-Washington corridor and the beltway highways that surround each city. Some of the speakers were healthcare researchers and biostatisticians who worked for major research institutes like Johns Hopkins University Medical School and the National Institutes of Health, and Dr. Hennessey himself, who talked about his work at the Social Security Administration. The range of potential real-world industrial applications of mathematics, operations research, and statistics was enough to make me dizzy with the prospects of what I could do with my skills and capabilities. What was most encouraging was that the speakers were all incredibly intelligent, equally articulate, and highly credible business professionals who understood not only how to apply mathematics but also the business world in which each respectively operated. And, in many ways, they were just like me and had just completed the same math degree that I was now undertaking. I felt confident that this is what I was meant to do.

Suffice it to say, I was blessed with several experiences and opportunities in my undergraduate years at Loyola that without a doubt put me on the right track in my education and career path.

Dr. Hennessey pointing me toward operations research was the catalyst. My advisor, Dr. Rick Auer, a statistician (Ph.D. from Iowa State University), was not only a great (and tough) statistics professor but also became a close friend (we shared a love of baseball, went to more than a few Orioles games together, and shared a few beers in Fells Point, a popular destination for the yuppie and college student crowds). Rick introduced me to biostatistics as a potential career field, which I seriously considered before realizing that, given my personality and interests, I was much better suited to O.R. in a business or corporate setting. After many conversations, Rick finally agreed.

In my junior year at Loyola (and continuing my senior year), a family friend helped me land a part-time job/internship/apprenticeship that provided me with great early-career experience and the opportunity to see operations research at work in the real world. It also helped steer my career direction and choices with greater clarity. That friend was Joe Burk, whom I knew from my family's church. He earned a B.S. in mechanical engineering and an M.S. in operations research from Johns Hopkins University and worked his entire career at a boutique niche DoD contractor in Towson, Maryland, that specialized in building mathematical models for analyzing a variety of military phenomena, e.g., predicting whether a Patriot missile will hit its target. Joe's specialty was a large FORTRAN program (known as COVART – *Computation **O**f **V**ulnerable **A**reas and **R**epair **T**imes*) that simulated kinetic threats being fired against an aircraft (e.g., rotary- and fixed-wing) and predicting P_k or **p**robability of **k**ill (i.e., the probability of defeating the aircraft by disabling any combination of systems, e.g., hydraulics and propulsion). Joe's firm had contracts with the U.S. Army, Navy, and Air Force to apply COVART to a variety of military aircraft. COVART was originally written in the FORTRAN IV computer language and required an upgrade to FORTRAN 77 and an extensive rewrite to leverage top-down, structured programming techniques and increase the readability, understandability, and usability of the model by third-party partners and customers. I was hired as an intern programmer/analyst, working two days per week during the school year and full time in the summer, to execute the software upgrade and rewrite. (I also worked on other small programming and data processing projects, as demand dictated.) Although the work was a bit boring, the experience of working on a programming project of that magnitude was invaluable.

Beyond the key learnings gleaned from my primary day-to-day job working on real problems and real systems in the real world, there were *three* ancillary, but critical, benefits of working for Joe. First, his company was replete with a dozen mathematicians, statisticians, and operations researchers who all had multiple graduate degrees and were more than happy to share information about the projects and models that they were working on, as well as help me with my math homework problems that were of more than modest rigor! For example, Dr. Bob Bennett (Ph.D. in electrical engineering from Johns Hopkins University), who ran the company's Towson office, had completed his Ph.D. dissertation on *signal processing analysis* to detect and identify radar signal "signatures" and how they change under different conditions. He's a super smart guy with a quirky sense of humor. Strangely enough, he liked to *type* on a typewriter (go figure), which I hated, so I paid him to type my term papers – lunch money, he called it!

Second, Joe's office had a library with a complete set of journals, including *Operations Research* and *Interfaces* (now the *INFORMS Journal on Applied Analytics*), that I read during lunch and could take home whenever I wanted. The journals, which were full of problems, models, and real-world case studies, enabled me to gain *invaluable* exposure to the state-of-the-art theoretical research and practical real-world industry applications of O.R. These articles provided me with a foundation of knowledge and understanding, well beyond what I was learning in class, that further incentivized me to pursue a career in operations research (which has since evolved to include analytics, data science, and AI).

Lastly, an invaluable lesson that I took away from working for Joe on DoD projects for two years convinced me that once I graduated, I no longer wanted to work on federal government contracts, despite the solid job security, but rather in corporate America, where capitalism was the order of the day and not quite as much of the "red tape" and bureaucracy of government contracting.

As I progressed through my junior year coursework and focused on operations research as my career direction, my thoughts turned to postgraduate employment opportunities – and then, I got a wake-up call from Drs. Hennessey and Auer. They informed me that to get a job and work in operations research in corporate America, I would most certainly need a *master's* degree. They said that a bachelor's degree would get me a job as a "programmer/analyst," but a master's degree would be *mandatory* to get a "seat at the table" where the business problems would be analyzed and models formulated. (The "fun" part! Not just the coding!)

Armed with that insightful advice, I set out to research graduate school programs in operations research. I quite literally wrote to *50+* universities to obtain literature on their O.R. master's degree programs. (I later donated all of the materials to the Loyola Math Department library for future students to use!) After my exhaustive research, I narrowed my choices to a handful of the top schools in the country (ranked by my preference):

1. Georgia Institute of Technology, School of Industrial and Systems Engineering (ISyE)
2. Purdue University, School of Industrial Engineering
3. University of Michigan, School of Industrial and Operations Engineering
4. Penn State University, School of Industrial Engineering
5. George Washington University, Operations Research

My decision became easier and more readily apparent when only Georgia Tech and Purdue offered me "full-ride" research assistantship scholarships that paid both tuition and a modest stipend to cover my living expenses. Based on several criteria, I chose Georgia Tech, which was another pivotal event for me in my career path trajectory:

1. No thesis required (I had zero interest in writing a thesis).
2. 48 hours of coursework (16 three-credit courses covering a variety of disciplines and topics).
3. M.S. could be completed in one calendar year (Georgia Tech was on academic quarters at the time).
4. Focus on O.R. as opposed to IE-type courses (four three-credit courses per quarter).

5. Georgia Tech is located in Atlanta, and Purdue is located in West Lafayette, Indiana; being a "city boy" and having visited Atlanta on vacation and for my campus visit, I felt quite a bit more comfortable and "at home" in a city versus a campus set among corn fields.

As it turned out, I made the right decision for multiple reasons. Georgia Tech's Graduate School of Industrial and Systems Engineering has been consistently ranked No. 1 by *U.S. News & World Report's* survey of Best Graduate Schools in Engineering since 1990. (I wouldn't have gone wrong going to Purdue as they are ranked No. 2). My first employer out of tech, American Airlines, had already started interviewing and hiring M.S. in operations research graduates (from both Georgia Tech and Purdue, as it turned out) for their growing O.R. Department; I was the second new hire of what turned out to be a cohort of more than 50 Georgia Tech ISyE graduates in O.R., IE, and statistics!

When I got to Georgia Tech ISyE in 1986, I realized how fortunate I was to be there. The O.R. faculty was literally and figuratively "world-class" award-winning academic researchers. George Nemhauser, Ellis Johnson, John Jarvis, Don Ratliff, and Dave Goldsman, just to name a few of many, were all top-notch academic researchers in their respective fields, but were also *practitioners* and *entrepreneurs* who started companies to apply their expertise and business acumen. [Jarvis and Ratliff started, grew, and eventually sold **CAPS** Logistics (**C**omputer-**A**ided **P**lanning and **S**cheduling) to Baan, an enterprise resource planning (ERP) software company. Nemhauser started Sports Scheduling, LLC, which was contracted for several years with Major League Baseball to generate the MLB season schedules. Jerry Banks and John Carson started their own simulation analysis company and completed projects for Coca-Cola, among others.]

The Georgia Tech student body represented the very best intellects from around the world, including the United States, Latin America, China, Eastern Europe, and India. My roommate Chris Hane earned a Ph.D. and is now a VP at Optum where he leads healthcare-related AI research. Ananth Iyer, Ph.D., served as a Department Chair and Professor of Operations Management at Purdue Krannert School of Management and is now Dean of the University at Buffalo School of Management. The list of accomplished alumni goes on and on.

The Georgia Tech M.S. in operations research program was simply a perfect fit for me and my interests. I had a strong undergraduate math degree focused on linear algebra, statistics, and operations research, which provided the ideal foundation for my 16 courses in theory and applications:

- Linear, Integer, and Nonlinear Programming and Decomposition Methods
- Probability Models, Probability Theory, Queueing Theory, and Decision Theory

- Regression Analysis and Time-Series Analysis
- Production, Inventory, and Distribution Systems
- Engineering Economy and Replacement Analysis
- Simulation Analysis

While attending classes, I earned my scholarship funding as a research assistant working for Professor Dave Goldsman. Dave was conducting theoretical research on selecting the statistical population with the largest mean using simulation analysis and a variety of selection sampling techniques. I found myself once again in the FORTRAN simulation modeling business. The added experience of writing code and analyzing data for a living was invaluable. Little did I know, these methods would come in handy later on at the beginning of my career.

Nearing graduation and ready to join the workforce, I interviewed with multiple companies, including AT&T Bell Labs (New Jersey), Stone & Webster (Georgia), and BDM (Virginia). My dad spent his entire career with AT&T (Western Electric, Lucent), so I thought that was worth a look. Stone & Webster wanted to send me to the Middle East to work on managing a dam construction project. BDM was another DoD Beltway Bandit and wanted me to work on its FORTRAN simulation model called **R**apid **R**unway **R**epair (**R**^{3}) that estimated repair times for runways damaged by various types of kinetic bombs. Been there, done that with COVART – check please!

I heard through the grad student grapevine at Georgia Tech that *American Airlines (AA)* was aggressively interviewing and hiring M.S. O.R. graduates for their burgeoning O.R. group. I had read about AA O.R. in multiple issues of the *Interfaces* journal, which featured several airline O.R. case studies. An airline seemed like an ideal environment for applying O.R. with expensive assets, such as airplanes, hangars, and gates, and considerable expenses, like crew and fuel costs, to manage. Again, relationships proved key for me to gain an introduction to AA O.R. Fellow Loyola alum John Leimkuhler, known as one of the "star students" of the Math Department, had gone on to earn his M.S. in O.R. from Purdue, where his uncle, Dr. Ferdinand Leimkuhler, was the department chairman. John was already working at AA O.R. at their headquarters in Fort Worth, Texas. I knew John, but he didn't know me because I was only a sophomore when he graduated from Loyola. I decided to send him a cover letter and copy of my resume, leveraging the Loyola Math Department connection, and he passed it along to the hiring managers and put in a good word for me. Fortunately, I was called for an interview, and after successfully navigating a very long, full-day gauntlet of interviews at AA, I was offered a position as an O.R. analyst at the AA HQ in Fort Worth.

Timing, as they say, is everything. As they also say, better to be lucky than smart – ideally you are both. I was the 40th person hired into the AA O.R. group in October 1987 and the *second* from Georgia Tech ISyE. My timing was *perfect* to join AA O.R., and I consider myself very lucky to have been hired there. The group was pioneering airline operations research with innovations in pricing and yield (revenue) management, flight scheduling, crew scheduling, spare parts inventory management, and airport air and ground operations simulation analysis. The group was *so* successful that AA senior management decided to form American Airlines Decision Technologies (AADT), a wholly owned subsidiary of American Airlines, to offer O.R.-based consulting services to other airlines and travel industry-related entities. Later, in 1993, AADT merged with Sabre Development Services to form Sabre Decision Technologies, which later became Sabre Airline Solutions, the world's largest (by overall market share) exclusive airline O.R.-based software products and consulting services company. (It was later purchased by CAE in March 2022.)

Flying High (and Overseas) at AADT

I was hired by Mike Parks, one of the top three executives at AADT, as an operations research analyst/consultant and assigned to an airport simulation analysis group that was applying the FAA's recently released discrete-event simulation model called SIMMOD (zero points for naming creativity) to analyze and evaluate airport airspace and ground operations and design alternatives (e.g., airspace configurations and new runways) to efficiently utilize and maximize airport throughput capacity. Initially, I was assigned to work on AA flight schedule-airport operations expansion projects at Raleigh-Durham and Nashville Airports using SIMMOD. I spent *a lot* of time in **a**ir **t**raffic **c**ontrol (ATC) towers and **T**erminal **R**adar **A**pproach **Con**trol (TRACON) working with air traffic controllers and learning how airports work, i.e., processes, policies, procedures, communication protocols, terminology, and a host of acronyms. *(Lesson No. 1: You cannot model any phenomena accurately on a computer unless you thoroughly understand how it works in reality.)*

At this early point in my career as an O.R. analyst on a project, I was just beginning to learn about analyzing and modeling complex, real-world phenomena such as airports using data and sophisticated models like discrete-event simulation. My Simulation Analysis class at Georgia Tech went a long way to prepare me for the "basics" of modeling systems, including unloading ships at a harbor or customer queues waiting for a bank teller. However, capturing all of the data necessary to build a model of an *airport* and simulating traffic loads running through multiple airspace and airport runway/taxiway configurations to determine which ones were most efficient/effective at minimizing air and ground delays was a *completely different level of modeling sophistication and complexity.*

In January 1988, AADT secured a consulting project with the Swedish Civil Aviation Authority (CAA) to evaluate multiple new airspace configurations at Arlanda Airport in Stockholm, Sweden, to determine (using discrete-event simulation analysis/SIMMOD) which one(s) would produce minimal delays. I was assigned as the analyst and learned several lessons after completing the project:

- Show up, diligently get your work done, and underpromise and overdeliver.
- Don't complain... ever, about anything.
- It is very difficult to communicate complex business subjects with folks for whom English is a second language (and shame on me, I didn't speak Swedish).
- How dark it is (no sun for two weeks) and how wet and cold it is in January in Sweden.
- Make sure you get *as much as sleep as possible* on the flight from the United States to Europe, because if you don't, you will *never* recover from jet lag (I had to take a nap every day at lunch!).

Despite the severe jet lag, extreme darkness, dankness, and cold (and a language barrier), my colleague and I successfully completed the project and were able to help the Swedish CAA select the minimal delay airspace configuration. In hindsight, we probably could have picked the right airspace configuration just by visual inspection of the designs (one had a severe built-in bottleneck), but the simulation quantitatively ensured the right decision. [I was even able to use my Georgia Tech graduate research on selecting the population (airspace design) with the minimum mean (delay) and published and presented a paper with Dave Goldsman at an INFORMS conference!]

After Stockholm, I was off to Spain in the spring of 1988. AADT secured two consulting projects with the Spanish Aviation Safety and Security Agency (CAA): (1) Analyze operations at Madrid-Barajas Airport, with emphasis on factoring in the effect of military air traffic at adjacent Spanish Air Force Base Torrejón (a lot of F-16 traffic) and (2) analyze the impact on ground (gating and taxiway) operations of a new, longer runway being built at Palma de Mallorca Airport off the coast of Spain. I went to Spain to help out on the Madrid project, but was assigned primarily to Palma. Although it was fun to visit two great Spanish cities and enjoy some phenomenal cuisine, neither of these projects presented any real complex technical challenges, but I did learn a lot about project management, people management, and client handling from my project manager, Jim Crites.

Jim was a retired U.S. Marine Corps Major (and later a USMC Reserves Lt. Col.) who went through ROTC at the University of Illinois Urbana-Champaign while earning a B.S. in business administration and earned an

M.S. in O.R. from the Naval Postgraduate School in Monterey, California. After AADT, Jim worked in corporate real estate, overseeing airport projects at AA, and later served as EVP of Operations for DFW International Airport, overseeing numerous facility expansion projects, including inter-terminal passenger railways and new terminals. Jim was recognized by the President of the United States as a Champion of Change for the Transportation Industry in 2013. Everyone should be so lucky to learn from a PM like Jim Crites early in their career.

Being a Marine officer, Jim was, as one might imagine, a "buttoned up, squared away, shined shoes, crisp suit, and tightly knotted tie" kind of fellow. I paid close attention to Jim's appearance, mannerisms, the way he carried himself, and how he interacted with clients and their "handlers" and any intermediaries that might confound a project's objectives. In addition to being "book smart" and technically astute in all phases of the "job," Jim was "people smart," politically savvy, and always looking at each angle in every situation. Jim was calm under pressure and knew exactly when to play it cool and when to draw the line; he knew when to show just enough strength in his tone, posture, and body language to make sure everyone knew he was stalwart. I never worked on a project with Jim Crites that went sideways – we always got the job done.

As a Marine officer, Jim *really knew* how to handle and manage people and was well-known for both his stern, candid feedback when you needed to improve and his good-natured ribbing if you ever got too big for your britches.

After the Madrid and Palma projects, I got to continue working with Jim Crites on various small projects involving analyzing AA's schedule and capacity expansion of DFW and examining the impact of high-speed taxiway exits to shorten taxi times.

Then, came the opportunity that immediately set my career on a steeper trajectory. In 1989, after contributing as an analyst on several SIMMOD projects in the United States and abroad, I was given the opportunity to run not one, but two of my own projects.

AADT secured a consulting project with the O.R. Department of Qantas Airways, working hand in hand with the Australian Civil Aviation Safety Authority (CASA) to analyze the impact of a newly extended heavy aircraft long runway at Sydney Kingsford Smith Airport, in Sydney, New South Wales, Australia. I was still *way* down the pecking order of seniority in the SIMMOD group, but none of the more senior project managers were excited about the prospect of the long (24-hour) journey to Sydney (via a connection in Los Angeles and a fueling "tech stop" in Papeete, Tahiti) and two weeks of temporary duty "down under." (They all had families, including Jim Crites, that they didn't want to be away from for the time of the project.) When my boss Mike Parks said he needed someone to manage, lead, and execute the project, I (half-joking and simultaneously timid and brazen) said, "I'll do it" – my smartest career decision to that point. *(Lesson #2: Whenever opportunities present themselves, especially early in your career when you are trying to make your mark, say "yes" far more often than you say "no" because the "big games" don't come around all that often.)*

Mike asked whether I was sure and said I could do it, but not to "f@&k it up." Mike always knew how to instill confidence in me, much like Gen. George S. Patton Jr., who used profanity for effect, and it worked!

As the old proverb says, *"Be careful what you wish for, you may get it."*

Data Down Under

The data collection effort in the Sydney project was *massive*, and the airport data modeling effort presented a few challenges that required quite a bit of creativity to capture. We had to "trick" SIMMOD into letting us do what we needed. One scenario in particular, among many others, bears explanation.

A very interesting feature of the newly extended "long runway" is that although it was intended for heavy aircraft, e.g., Boeing 747s, when much *lighter* aircraft (think Cessna propeller plane) landed, they were able to *stop short* of the intersecting runway, such that air traffic control regularly permits *simultaneous arrivals/landings on both of these intersecting runways.*

Well, SIMMOD most certainly did not under any circumstances understand or allow aircraft to land simultaneously on intersecting runways! I had the bright idea to create a *phantom runway* (i.e., adjacent to the long runway that *did not intersect* with the crossing runway) that only allowed *lighter* aircraft to land, as long as there were no other aircraft landing on the adjacent parallel long heavy aircraft runway. *Get the picture?*

It's been famously said, *"The English and the Americans are separated by a common language,"* which made the project even more fun interpreting words like "bitumen" or asphalt on the airport tarmac and phrases like "the custard's come off the plate," meaning things went awry. This made complex communication about airport operations even more entertaining!

At the relatively young professional age of 25, the opportunity for me to manage a high-profile external client project overseas involving a major foreign government entity (i.e., Australia's FAA equivalent) that would directly and significantly influence the way in which their big city airport operates was nothing short of a phenomenal learning experience and a chance to prove what I could do. As project manager, I was *100% responsible and accountable* for all aspects of the project delivery: project planning and execution with respect to scope, timing, budget, expenses, resources, deadlines, deliverables, quality (technical and presentations), status reporting, problem resolution, billing, payment, doing the work, leading and managing the work, and managing the client customer and stakeholder relationships. While it was quite stressful at the time, it was an *invaluable "trial by fire" learning, career development, and advancement experience.* By successfully completing the project, I was able to prove to my superiors, and the client, that I could be trusted and deliver a project from beginning to end on time, within budget, and with *100% customer satisfaction on scope, quality, and value, as promised.* This was the first major building block of my career at AADT.

There is a tripartite piece of advice that is apropos here for young professionals. It is advice that I received early in my career that was invaluable in my career development and advancement:

- Say *"yes"* more often than you say *"no"* when asked to participate. (I could have turned down the opportunity to manage the Sydney project and let whatever trepidation scare me into oblivion, but I said "yes," and it made all the difference career-wise.)
- *Make yourself indispensable* – you want to be the *"go-to"* person when leadership needs something important done well.
- *Endear yourself to people* – be someone other people want to be around and have around, call upon, and follow (primarily your leadership, clients, and co-workers who will need your help to get things done).

We did such a great job with the SIMMOD project in Sydney that Qantas' O.R. leadership (and the CASA) asked us to stay in Australia for an *additional two weeks* to do another similar type of project at the airport in Melbourne. That is the *best customer feedback you can ever receive – when the customer asks you to perform more work before you have even completed the first project!* I don't even recall the scope or ask in that project, but we completed it successfully as well and enjoyed the heck out of the food, sights, and culture in Melbourne!

Movin' On Up

Typically, practitioners begin their career going through a progression. First, they run models built by others (that was my time on SIMMOD), modeling different scenarios through data. Then, they tweak, modify, extend, enhance, and maintain models built by others. Then, they design, build, and deploy their own models from scratch. The progression represents an increase in responsibility, accountability, and risk management. Then, they start managing others on individual projects and start managing one or more larger programs, each made up of one or more projects. Along the way, the size of the teams, organizations, budgets, and value targets they are managing continues to grow. My career followed this exact progression.

After my time on the airport analysis team (SIMMOD), during which I was promoted to senior consultant from consultant, I was assigned to work on a program in AA's systems operations control (SOC). SOC is the "command center" for an airline, where FAA-licensed dispatchers are responsible for critical information and decisions, including aircraft weight and balance, fuel loads, weather, and safety. They also make mission-critical decisions about which flights to delay/cancel and how best to utilize available aircraft. For example, if one aircraft has a maintenance problem and another available aircraft doesn't depart for some time, a dispatcher will "swap" the aircraft, so both flights can depart with little or no delay. About one-third to one-half of an airline's flights on a daily basis will be affected by some type of operational impact or change.

Dr. Alberto Vasquez, a brilliant mathematician and expert in modeling scheduling problems, was very successful in building and implementing a model for SOC to optimize the assignment of arriving AA flights to FAA arrival "slots," i.e., air traffic control time windows when flights are permitted to arrive at a busy airport like DFW. In real time, the **a**rrival **s**lot **a**llocation **m**odel (**ASAM**) gathered data from AA's **f**light **o**perating **s**ystem (**FOS**) and evaluated all arriving flights at a given airport, some on time and some delayed, considered all available arrival slots, and then assigned all of the flights to arrival slots such that total flight arrival delays were minimized. ASAM was so successful that the solution was a finalist for the prestigious INFORMS Franz Edelman Award for best practical applications of operations research.

The next phase of the program was to build on the success of ASAM and create a new and improved, more robust and broader scope model, i.e., **m**odel for **i**rregular **o**perations (**MIO**), that would endeavor to optimize flight delays, cancellations, and aircraft swaps *across the entire AA flight network* to minimize total delays and passenger and flight disruptions. A far, far more complex problem to solve in terms of the massive amounts of real-time data required and the decision-making affecting a highly interconnected flight network with thousands of flights per day.

Long story short, we were not successful in building and deploying MIO. Alberto and I wrote down a very elegant, very comprehensive mathematical model to solve the irregular operations problem. And even though all of the data we needed was stored in FOS, getting the data into one place for populating the model to solve the problem proved to be beyond our budget and capabilities. More importantly, all of these types of mission-critical flight delay and cancellation and aircraft assignment decisions were traditionally made *manually* by human dispatchers who (including their leadership) were not prepared to consider giving up that kind of control to a "model" or "computer" – notwithstanding the extraordinary success and effectiveness of ASAM. After several months of meetings that led nowhere, the project was canceled.

Someone once said, "When a door closes, a window opens."

That was certainly the case for me personally after the SOC MIO program ended, in effect, a failure, due to some of the usual suspects, i.e., data issues, problem scope and complexity, change management, budget, and a dose of politics.

In the early fall of 1990, Mike Parks assigned me to work on a project with AA maintenance and engineering long-range planning (LRP) at AA's primary heavy maintenance base in Tulsa, Oklahoma. As AA's fleet had grown from 200 to 600 aircraft, the process of planning for and scheduling heavy maintenance "check" (or "overhaul") activity and hangar capacity was becoming more complex than the LRP team could handle with large sheets of paper tacked up on the wall marked with colored pencils (the historical incumbent solution) or even Microsoft Excel (the newfangled spreadsheet macro for scheduling the 250 narrow-body aircraft fleet of MD-80s was taking *12 hours to run* and oftentimes crashing before reaching a solution!). Senior leadership was growing more and more concerned about ensuring completion of all heavy maintenance checks on time (avoiding fines and grounding of aircraft), minimizing costs within FAA rules, and not running out of much needed hangar capacity. No one in LRP had solid solutions to any of these problems.

Heavy maintenance checks are required to be completed on each aircraft periodically, the time window of which is governed by a designated flight hour limit (e.g., every 15,000 flight hours) and cost, at the time, $1 million per aircraft check (or ~$2.4 million in 2023). The objective is to bring an aircraft in for maintenance *as close to the flight hour limit as possible without going over* to maximize the "yield" of the check. If the flight hour limit is *exceeded*, then FAA fines and penalties are incurred and the aircraft is grounded. If the aircraft is brought in *too early, well before* reaching the limit, then more checks than are legally required are completed over the life of the aircraft, thereby unnecessarily increasing maintenance costs.

In roughly six months of research, analysis, and iterative design, development, and testing, I was able to successfully model and solve the problem using a *job scheduling on parallel machines with fixed job deadlines* modeling approach solved with a "greedy heuristic" algorithm. My colleague Bobby Johns (one of the best software engineers I have ever worked with) developed a colorized Gantt chart user interface that digitally emulated the large sheets of paper and colored pencil approach on an Apple Macintosh PC (IIx running a Motorola 68000 chip). The industrial engineer on the project (who came up with the original concept) estimated that the model would increase check yields to nearly 100% and, on the 227 widebody aircraft fleet alone, would eliminate 1–2 heavy checks over the life of the fleet or a cost avoidance of $227 million to $454 million (or ~$545 million to $1.09 billion in 2023)!

This project was a tremendous success and literally changed the trajectory of my career, given the business value and economic impact that our automated, intelligent solution created (i.e., hundreds of millions of dollars in maintenance costs avoided over the life of the fleet and an automated tool that analysts could use to run all manner of planning scenarios in a matter of minutes on their desktop Apple Macintosh PC that used to take hours, days, or weeks, depending on the scope).

The project and all of the lessons learned are detailed in two articles later in the book:

- Airworthy," *OR/MS Today*, December 1992.
- Broaden Perspective: Consulting concepts come to life for author while working on American Airlines maintenance project," *OR/MS Today*, December 1993.

The learning experience on the project was *foundational* for me and established a set of principles that guided my career in analytical sciences. Three quotes from O.R. luminary Dr. R. E. D. "Gene" Woolsey (Colorado School of Mines professor and industry consultant) summarize my key learnings:

1. *"If you try to tell someone how to do something better or differently without understanding how they do the job today, then you are a fraud."*
2. *"Finding the 'optimal' solution is often not nearly as important as putting the solution values into a form that the client is accustomed to seeing."*
3. *"A manager will prefer and opt to live with a problem that they cannot solve rather than implement a solution they cannot understand."*

I spent several weeks in Tulsa sitting next to the LRP planners learning everything there was to know about heavy maintenance check activity and facility capacity planning and scheduling *before* I started modeling the problem. The model and system we built were simply a more formalized, optimized, and automated version of the business process flow that the planners were already executing, albeit their way was much less efficient and more time-consuming. Therefore, it was relatively easy for the planners to make the (not so big) leap to use the new computer model/system, because they intuitively understood how it worked and saw it as a "big calculator." Mission accomplished!

Bobby and I, along with a couple of additional engineers, spent another year or so enhancing and extending the model to accommodate different planning scenarios, such as third-party aircraft maintenance (for profit), FAA directive-related checks, and landing gears (we achieved 99% yield based on cycles, i.e., one cycle equals one takeoff and one landing, together), and rewriting the C code into C++ to leverage the benefits of object-oriented programming.

As a result of the above success, in 1992, I was promoted to principal (i.e., manager), which was the first level of management at AADT. I worked on a series of smaller projects, including forecasting aircraft engine removals and engine repair shop arrival flow (using the binomial and Poisson distributions). (The engine shop needed a business case for expansion to handle increased demand due to fleet growth, but they needed a more scientific basis for estimating how many more engines would be arriving at the shop for unscheduled repairs and scheduled maintenance.) I also did some consulting for other airlines around the world using our maintenance planning and scheduling tool. My team and I also prototyped a model for scheduling tasks (e.g., inspections and repairs) and resources (e.g., people, tools, and equipment) on overhaul checks *inside* the hangar, but the interest simply was not there from the business folks at any client company to implement such a solution.

In 1993, Mike Parks asked me to take responsibility for all of the models and systems that scheduled pilot training activities (i.e., new hires, recurrent, and check rides) and facilities (i.e., instructors, classrooms, and simulators) for the AA Flight Academy. These were optimization-based scheduling models designed to ensure that there were no gaps in crew training requirements and mission availability. In addition, I inherited the crew manpower planning system that utilized a multiyear time horizon integer programming model to determine the optimal number of pilots and flight attendants to hire each month, factoring in fleet growth, changes in aircraft type fleet composition, and both aircraft and crew retirements, to ensure adequate line and reserve crew resources. We employed this model on behalf of AA and other U.S. airlines. Although these systems were all mission-critical, there was not a lot of new development going on, so this work was largely in "operations and maintenance" mode.

In 1994, AADT merged with Sabre Development Services (the software arm of Sabre) to form Sabre Decision Technologies. I was promoted to director that year in August, and I added responsibility for another set of systems that operated the Flight Academy. Merging two different cultures together, ours entrepreneurial and innovative, with another that was quite stodgy, bureaucratic, and resistant to change, was the biggest eye-opener for me. That said, I was also able to learn and grow by taking on some large, multi-million-dollar travel distribution system integration projects, e.g., merging Sabre Air Booking with a large tour company's booking systems – no analytics, but a good learning experience for managing large-scale system development/integration projects.

In July 1995, all of my crew-related programs were merged into a large crew system suite that was managed by another leader. At that point, my career took a much different direction, away from a focus on analytical sciences and into internet e-commerce and software product engineering companies, including startups.

(Chapter 3 on "Digital Transformation" covers a lot of the work that I did from 1995 to 2009.)

In summary, in 1995, the internet and World Wide Web were in their infancy with companies like Netscape (the first commercial web browser) going public. E-commerce was the next big thing – monetizing the internet by marketing and selling products and services online.

I saw firsthand in my previous job the business opportunity presented by Sabre and travel distribution. Mike Parks asked whether I wanted to lead the program to put Sabre's travel reservation capability on the internet and make it available directly to consumers to book their own travel online. Frankly, no one else wanted the job because they thought the internet was a "fad" that would be "gone soon" and that consumer-direct travel distribution was undermining Sabre's travel agency "bread and butter" legacy business model. People at Sabre actually said things like *"Why would a consumer ever want to book their own travel on a website?"* In hindsight, pretty ridiculous, right? Mike, who hired me, had been my boss and mentor for eight years and pretty much taught me everything I knew about business to that point; I was fiercely loyal to him. He needed my help, so I said, "Yes!"

Delivering Value

Travelocity, Sabre's consumer-direct travel distribution model (now part of Expedia), was the world's first completely automated real-time, internet e-commerce travel booking tool. For two years (1995–1997), I led the team that designed, engineered, developed, launched, operated, supported, and maintained the software technology platform underlying Travelocity, as well as functioned as the CTO in charge of hardware and network communication engineering and operations. A core team of four, which eventually grew to 75, took Travelocity from a "whiteboard drawing" in July 1995 to a successful launch at the CyberCafe in Soho, New York City, on March 12, 1996. I was there when our CEO, Terry Jones, made the *world's first airline booking in real time via the internet.* From there, we successfully licensed, customized, and hosted that software platform to become the internet travel booking engine for American Airlines, Sabre Web Reservations, Cheap Tickets International, Rosenbluth Travel, Canadian Airlines, and many other e-commerce travel distribution properties.

I realize this is a book about timeless lessons learned during a career in the *analytical sciences*, and Travelocity had little (OK, nothing) to do with analytics (other than some rudimentary AI laboratory experiments to automate booking steps from natural language queries – see Chapter 3 "Digital Transformation").

What my Travelocity experience *did* teach me was *invaluable* later in my analytics career (i.e., *building and deploying large, highly scalable, transactional enterprise systems*):

- *N-tier architectures*, separating the user interface (UI) from business logic and services from backend data services coupled by application programming interfaces (APIs) and messaging services (service-oriented architecture is the primary architectural design pattern of all systems today).

- *Scalability*, building software systems that can handle more traffic and users dynamically just by adding hardware capacity (this concept is the foundation of cloud computing).
- *Managing large program teams* of multidisciplinary engineers and analysts that span UI, DB, software, hardware, and networking, prioritizing competing requests with a limited fixed budget, and dealing with deadlines that do not move under any circumstances.
- *Being, in effect, the CTO* of a startup company and the "one throat for the CEO to choke" when things went awry.

Analytical sciences is not only about building *models* but also about putting those models into *production* and *embedding* them within *mission-critical enterprise business processes and systems that deliver business value and economic impact.* Tom Davenport, author of *Competing on Analytics*, said, *"Models make the enterprise smarter, but models embedded in processes and systems make the enterprise more economically efficient." That*, my friends, is the *end game*, your *raison d'etre*, and *that* is why you need to learn about building and deploying models as *microservices* embedded in larger enterprise system architectures and ecosystems if you are going to deliver value at scale with analytical sciences in corporate America, government, or the military.

After an incredible 10-year learning experience that launched my career, I parted ways with Sabre when a major reorganization left me with an uncertain path forward (I was most likely within six months of a promotion to VP, but now that was in jeopardy). Mike Parks left the company soon after I did and moved to the West Coast to work in Silicon Valley. I chose to go east to work as VP and CTO for a startup e-commerce solutions company in Tampa, Florida (a 33% increase in base pay plus bonus and equity participation was a big motivator). Sometimes, I regret the decision to not follow Mike, but the past is the past. Although I cashed out of the startup with a six-figure stock windfall, I often wonder how things might have gone had I followed Mike to the West Coast. This story would not be complete without acknowledging that Mike was the best manager, leader, "boss," and mentor that I ever had. He was also one of the best all-around business people with whom I have ever been associated. Mike recruited and hired me, trained and educated me about business, mentored me, and gave me ***all*** of the project opportunities that defined my career at AA/AADT/SDT/Sabre/Travelocity. Most importantly, he *understood* me and *valued* me for who I was, and he made me *better*, professionally and personally, for which I will be eternally grateful.

A key *lesson learned* here: *Never ever underestimate the inordinate value and political power of leaders near the top of the "food chain" with whom you have a close, trusted, mutually beneficial working relationship.* No one "climbs the corporate ladder" *alone* – while you are "climbing," your "sponsor" is "pulling you up the ladder by your shirt tail." Your performance is only one part of the equation. *Who knows you and wants you to be successful is far more important* in your promotion potential and to get the "plum" assignments. In the military, generals promote the majors and lieutenant colonels who they believe have the greatest potential to become a general one day. *The same principle exists in corporate America.* If you regularly change companies or organizations, *without continuous sponsorship,* it becomes more and more difficult to climb up the ladder because *you are no one's candidate* due to lack of tenure, loyalty, and trust that comes from years and years of working for the same leader/sponsor. Frankly, I took my success at AA/Sabre for granted as a result of my own performance, and I failed to really understand this principle until much, much later in my career, and by then, it was too late to advance to senior management.

With IPOs turning technology geeks into millionaires with stock options, I figured I would ride the "internet wave" for a while. I was able to parlay my Travelocity CTO experience into a series of CTO positions with e-commerce startups selling everything from computers to mortgages to Las Vegas hotel rooms online (none of which went public in an IPO, but several were bought out, and I did fairly well financially with bonuses and equity stake cash outs). In 2000, I launched a VC-backed startup of my own in the system integration management console SaaS software product market that ultimately got bought out by Cisco. I did not do much work in analytical sciences during this time other than some prototypes (see *Personal Mortgage Optimizer* in Chapter 3 "Digital Transformation").

In 2003–2004, I was recruited to help lead the turnaround of a division of McAfee (then Network Associates, based in Plano, Texas – north of Dallas) as VP of engineering and product development (i.e., IT Service/Help Desk product Magic Solutions, acquired by BMC Software), by leading a team of awesome software architects and engineers in re-architecting and reengineering the product and helping to drum up sales of new licenses and renewals with a revitalized product roadmap.

My experience at Network Associates was *invaluable* and taught me about the importance of "selling." My division GM, Jeff Honeycomb (another terrific boss), was a professional salesman, through and through, who began his career selling copiers *door-to-door, floor-to-floor* in the skyscrapers of Manhattan and worked his way up to selling computers and the "big money" of selling enterprise software product licenses. He had an engraved brass sign on his desk that said, *"Nothing happens until somebody sells something."* Truer words were never spoken. Not surprisingly, Jeff had me spend about 50% of my time in the field on customer sales and support calls as the VP of engineering and product development. New license deals and license renewals were *critical.* The fact was that the product was falling apart and customers hadn't had a new major release (x.0) or even a "dot" release (x.y.z) in *years* – they weren't happy *at all.* Jeff wanted me to hear the problems *firsthand* and see the challenges the customers and sales team faced, so I could fix it. Trust me, you don't have to get screamed at too many times in Dutch, German, French, and British English by bank presidents, military commanders, etc. (all customers), before you get *super motivated* to go and fix the @#$%&* software!

If you are going to lead an analytics organization, you will be responsible for "selling projects," and that means building relationships, understanding and dealing with customers and their problems, and creating value-based proposals on ways to fix them for which they are willing to pay you in the form of their scarce budgeted resources – precious in every company. "Top 10 Analytics Leadership Skills" in Chapter 11 identifies selling projects (generating demand) as the No. 2 *most important skill to develop.*

My time at Network Associates was a pretty good run professionally and financially (although no seven-figure "home runs"), and I grew a lot technically and managerially, adapting to run larger, faster-paced engineering organizations. That experience was *foundational* for designing, building, deploying, and operating large-scale, sophisticated enterprise systems, as well as running an engineering (or analytics) organization like a "business" (i.e., a profit and loss statement, which means "selling").

When the U.S. real estate market and the global economy crashed in 2008–2009, I was stuck in Scottsdale, Arizona, after working for a couple of startups, one that failed outright and both with *huge* CEO/board of directors shake ups (long, very sad story, don't ask). While I was struggling to find a job, I decided to start my own company offering virtual CTO services. I made a living through my LLC with a variety of e-commerce companies and even merger and acquisition due diligence projects. Then, I got a call from a former colleague who asked whether I was interested in working on a one-year contract for a healthcare insurance company as their acting director of analytics to design, develop, deploy, and operate an analytical data warehouse (ADW), a management console dashboard (to replace a 1-foot tall stack of reports), a suite of predictive analytics tools for enhancing patient risk diagnosis and outcomes, and conduct a few proof-of-concept (POC) R&D experiments against the analytical data. That sounded interesting for three reasons: (1) It was steady employment for a year; (2) as I mentioned earlier, at Loyola, I was interested in and seriously considered a career in biostatistics, a closely related field to healthcare analytics; and (3) it was a chance to get back to my roots. So, I decided to pivot back to my strong suit – analytics.

Every once in a while, the stars seem to align, and you have an opportunity to accomplish something really special that contributes to society, as well as to your client/company. A $2 billion healthcare insurance company in the Midwest (anonymized for confidentiality reasons) was well-known for its *award-winning* operational data warehouse (ODW) and highly functional data governance capabilities. The company operated *44 distinct enterprise systems* for handling claims, pharmacy, billing, payments, patient information, etc., and all of that data was integrated nightly for reconciliation in their ODW. The VP of data was a very savvy, very knowledgeable 18-year company veteran and was ready to invest *heavily* in an ADW and a suite of healthcare predictive analytics applications that would fill in the gap between the ODW and the company's actuarial department, which was exclusively focused on the mathematics of setting premiums.

When you are evaluating potential analytics projects, you are looking for a set of *attributes*. The healthcare insurance company project checked all of the boxes:

- *Customer/Client*: Data savvy with a proven track record of delivery data solutions with high-quality data platforms in place. Business savvy knows the company's relevant processes and systems inside and out. Rational, reasonable – realistic expectations about what it really takes and costs to deliver the type of solutions requested and willing to have thoughtful discussions regarding options and *pros and cons* of different approaches. Engaged – willing to be sufficiently hands on in the delivery execution process.
- *Budget*: Ample, more than sufficient to acquire the necessary software and hardware and pay for a full complement team of ~20 high-caliber professionals/consultants to design, develop, and deliver the solutions.
- *Timeline*: Ability to jointly set a realistic target to deliver the project in a timely manner (for this project, we chose one (1) elapsed year).
- *Scope*: ADW, separate and distinct from the ODW. A dashboard that enabled the CEO/COO to run the company off of an *iPad*; a set of 11 healthcare patient risk predictor models to replace incumbent expensive third-party software; and three POC projects to demonstrate *"the art of the possible"* regarding what could be done with data to tackle difficult problems experienced by the company.
- *Business Value and Economic Impact Potential*: Is this a project that will generate significant financial impact against the most relevant, important business unit key performance indicators (KPIs)?

As the acting director of analytics of this healthcare company (which I will refer to as HealthCo), I was basically the "quarterback" of the program. I called the shots and "ran the plays" for all things technical, managerial, and administrative. Because it was an outsourced consulting gig, I functioned as the engagement manager and was responsible for writing the proposals, contracts, and project plans; designing the solutions; recruiting the team of 20 professionals; billing and collections; and overseeing the delivery execution schedules and budgets for building the components. I personally managed the RFP processes for the three major technology platform components, as well as the three POC projects, because they were analytical and experimental in nature.

One critical hire was the *architect* of the new ADW and management console dashboard. He was a very accomplished, very experienced data and analytics professional with a lot of big company experience, and I asked him to basically run those two projects on his own because they were the largest and most complex components of the solution and were intertwined.

The RFP process ran very smoothly, and we were able to solicit bids from the top players in the market and then play them off of each other to get the best deal for the client. We ended up selecting name brands for the ADW and dashboard platforms, and the client wanted to continue to use a legacy statistical package as their analytics tool because they already paid for the licenses.

The 11 healthcare risk predictor models to predict the likelihood that patients would be susceptible to and develop maladies such as heart disease, diabetes, and various cancers were built using patient data and replaced expensive licensed third-party software (this action alone saved HealthCo *$2 million annually* in licensing fees).

The three POC projects were carried out in timely response to real, critical issues that HealthCo was facing:

1. Predict the likelihood that a patient would require spinal surgery (binary classifier): Spinal surgeries, and the associated costs, were spiking in HealthCo's service area due to an aging (and a not so physically fit) local policyholder population, heavy industrial workforce, and highly incentivized surgeons who wanted to try out the latest spinal surgical repair devices from large biotech manufacturing companies; skyrocketing spinal surgery-related costs were threatening HealthCo's financial performance if not checked.
2. Predict the likelihood that a patient would readmit to drug and alcohol rehabilitation after discharge (binary classifier): Resources were stretched thin at HealthCo's subsidiary that operated drug and alcohol rehab facilities, and readmissions were up because staff could not identify and properly tend to patients who were at the highest risk of relapsing.
3. Predict the likelihood that a patient would opt in for a new dental insurance product (binary classifier): HealthCo's marketing team wanted to "test the waters" for a new dental insurance product *before* investing heavily in sales and marketing campaigns.

All three of these POC projects were successful and provided invaluable insights into predictable patterns of demographic/psychographic/behavioral patient data that helped practitioners better allocate scarce resources and deliver better quality of patient care and better patient outcomes, more economically. *All* of the credit goes to the four professionals I recruited who did all of the heavy lifting: two healthcare economists who taught at a local university; a data science colleague who I previously worked with; and a longtime O.R. colleague from AA who had established her own consulting company doing this kind of marketing science work. Customer satisfaction was *through the roof* on these three projects, and the end product provided even greater justification and testament to HealthCo's investment in the power of analytics applied to voluminous healthcare patient data.

Client confidentiality limits what I can share, but suffice it to say, this project was successfully completed with a robust, performant ADW and dashboard, predictor models, and fresh insights into solutions for challenging business problems at hand. The solution was *so* successful that HealthCo offered the use of their new platform on a Data Analytics-as-a-Service (DAaaS) basis to other healthcare insurers. The DAaaS program was also so successful that the DAaaS platform was spun off as a separate company from HealthCo and was purchased by another healthcare informatics company. HealthCo became a customer of the DAaaS platform they had built!

By design, this was a short-term engagement for me of slightly less than one calendar year. Despite the brevity, it was one of the most successful, and impactful, programs that I have run in my entire analytics career. HealthCo benefited from the investment in world-class products and solutions, and in turn, their patients and policyholders benefited from better quality of care and outcomes. A win-win-win!

After HealthCo, I continued working on "time for money" CTO consulting gigs through my LLC for a few more years, while looking for a career opportunity that I was passionate about and had some financial and professional upside. I reconnected with a former colleague and fellow classmate from my SMU MBA program on an email thread ironically intended to help another fellow SMU alum who was searching for a job. He was working for Southwest Airlines at the time, and he asked, *"Aren't you an airline O.R. guy?"* I replied in the affirmative. It turned out that Southwest was looking for a senior manager to run their 11-person O.R. group (called Optimization Solutions).

Given my background and connections, I got the job and moved with my family back to Dallas (the third time was a charm). I was pleased to find that a few super-talented former AADT colleagues were also on my new team at Southwest. The majority of the group was focused mainly on crew-related optimization solutions, i.e., flight and cabin crew schedule planning development, as well as all of the models that optimize crew training schedules. (A bit of *déjà vu* from my AADT days.) There was also a four-person team working on an R&D project with network operations control (NOC) to develop a real-time airline irregular operations optimizer to put flight schedules back together after major disruptions due to weather, air traffic control, etc., and minimize adverse effects of delays and cancellations, while ensuring passengers got where they were going (more *déjà vu* that made me shudder after my own failure with the model for irregular operations (MIO) at AA SOC). Fortunately, this team had quite a bit more intellectual horsepower, expertise, and experience (i.e., three Ph.D.s, including an INFORMS Franz Edelman Award winner who had designed, developed, and deployed ***SkySolver***, a crew irregular operations optimizer at Continental Airlines).

Over a period of roughly five years, I am proud to say that we *doubled* the size of the O.R. group while building on the success of the crew optimization team, achieving success with the NOC irregular operations optimizer (known at Southwest as **The Baker**, posthumously named after an NOC supervisor of dispatch, Mike Baker, who conceived of the idea for the solution), and expanding into some new areas, like jet fuel and liquor inventory optimization. The crew optimization team alone delivered $100 million annually in crew cost avoidance by optimizing pilot and flight attendant schedules. The NOC program team delivered The Baker, which helped to improve airline on-time performance overall *(by 2.11 percentage points, annually)* and significantly *(by a factor of 2X)* during irregular operations caused by major winter storms. The Baker became, and remains, an invaluable decision support tool for NOC supervisors of dispatch to better manage airline network disruptions and help ensure a better customer experience. After years of rigorous testing and validation, The Baker solves for and re-optimizes flight schedules in the wake of isolated network disruptions in *less than 5 minutes* and major network-wide disruptions in *~30 minutes*, a process that required *several hours when done manually*. The Baker won two prestigious industry awards, AGIFORS Operations Best Innovation Award and FICO's Decision Management Award, and multiple team members won the coveted President's Award, the most prestigious prize awarded to associates at Southwest Airlines.

[For more about The Baker Project, see Hagel, J., Brown, J.S., Wooll, M., & De Maar, A. (2018). "Southwest Airlines: Baker workgroup: Reducing disruption and delay to accelerate performance." *Deloitte Insights* (A Case Study in the Business Practice Redesign Series From the Deloitte Center for the Edge), 1–13.]

Jet fuel (JP5) represents an airline's second largest operating expense after crew. Together with one of the O.R. analysts on my team, I partnered with the jet fuel supply chain management department to build a new solution to more accurately forecast fuel demand, annually purchase fuel from suppliers under contract at minimum cost, and optimize fuel inventory levels at all 100 U.S. airports. (Fortunately, that analyst was brilliant and had a Ph.D. in supply chain O.R. from Clemson University; he now works for Amazon in Seattle.) Having an accurate jet fuel demand forecast and then managing jet fuel inventory levels at each airport are *critical* for an airline; if you buy more fuel than you actually need, then you have to pay to store it (holding costs), and if you don't buy enough in advance, you will have to pay a premium on the "spot market" or worse, cancel some flights (shortage costs). That solution, which was implemented in *less than one calendar year by that single O.R. analyst* (as a set of *microservices* integrated with enterprise jet fuel purchasing and management systems), delivered a total cost avoidance of *$38 million* over a three-year period and delivered optimized jet fuel decisions *in a matter of minutes*, replacing a more rudimentary spreadsheet-based solution. The solution was so compelling it won two awards: Alteryx's Best Business ROI Award and the Drexel LeBow Analytics 50.

Anytime you can reuse a performant, working solution to solve *another* similar problem, it is a *win* for the enterprise. Liquor is a commodity that must be managed similarly to jet fuel (Jack Daniels, among other brands, instead of

JP5 jet fuel). Inventory must be bought and stored until needed. You need to forecast demand by product, decide how much of each product to purchase in each airport catering station location, and then manage inventory levels and purchasing as demand fluctuates over time. *Sound familiar?* One O.R. analyst took the jet fuel optimization solution "as is," with no model changes, applied it to *liquor,* and realized a one-time *$12 million to $18 million* capital expenditure benefit through optimized liquor inventory decision-making *in less than one elapsed year.*

My experience at Southwest Airlines' Optimization Solutions *proved without a doubt* that a relatively small group of super-talented and committed analytics professionals can make a *huge* difference when allowed the freedom to innovate and execute and aim that robust capability at the problems with *big financial and economic multipliers,* like crew, jet fuel, liquor, and irregular operations. It was truly an honor and privilege to lead that team and witness the business value that they created and the economic impact that was manifested by their analytical expertise and passion to make things better!

During my tenure at Southwest, I also served as director for enterprise data, managing all of the data warehousing, enterprise reporting, and data science tools and dashboard platforms, projects, and people. We delivered some *huge* projects; two in particular come to mind. One is a **customer data warehouse** for marketing that was credited with enabling *$100 million* in incremental revenue annually by targeting the *right marketing campaigns and offers* at the *right customers,* at the *right time,* and at the *right price.* (The customer data warehouse won the Teradata EPIC Award.) The second is all of the data pipelines/extract, transfer, and load (ETL) code and data warehouse tables to enable the data flow from a new passenger reservation system to the myriad downstream enterprise applications that needed that data to function and execute business processes, such as credit frequent flyer miles, process refunds, and enable proper revenue accounting, among many others. That project required *more than 3,900 test cases* that had to pass for acceptance and *over 150 staff working for two years to complete – on time.* That was, *by far,* the largest and most impactful project team that I managed and led in my entire career. An *amazing* group of leaders, architects, engineers, and analysts who *never wavered* in their commitment to complete their part of the *largest system project in the history of the airline industry.*

After my time at Southwest Airlines came to a close, I was looking for an even *greater* challenge with an even *larger* company to ply my trade in analytics and data science. For me, all of my very best career opportunities have come from a *referral* from someone I already knew, like a colleague or recruiter. The next job was no different.

In 2019, my former FICO Xpress account executive at Southwest told me that Walmart Global Tech was aggressively hiring to dramatically expand its data science capabilities, particularly in the supply chain. He offered to submit my resume to the then chief data and analytics officer, with whom he had previously worked. After a lengthy and rigorous interview process, I was hired in January 2020 as director of data science, focused on U.S. supply chain-related domains.

Walmart operates the world's largest omnichannel supply chain stretching from foreign and domestic suppliers through a network of hundreds of fresh and ambient grocery and general merchandise consolidation centers, distribution centers, and fulfillment centers to 4,616 U.S. brick-and-mortar stores (supercenters and neighborhood markets) and 599 U.S. wholesale clubs (Sam's Club), all the way to the consumer's front door, refrigerator, and pantry. With a fleet of 10,000 tractors, 80,000 trailers, and 12,000 truck drivers traveling 1.1 billion miles every year, Walmart operates the largest private transportation fleet in the United States. Walmart continues to strive to *streamline, automate, and optimize* end-to-end fulfillment operations to make the supply chain more *economically* efficient. Walmart does this in many ways, including:

- Augmenting associate labor with advanced robotics systems.
- Experimenting with driverless delivery vehicles.
- Applying all manner of data science to optimize operations.
  - In 2023, Walmart was awarded the INFORMS Prize for best overall utilization of analytics on an enterprise scale and basis.
  - In 2023, Walmart was awarded the INFORMS Franz Edelman Award for the most economically impactful application of analytics.
    - Transportation network retail truck routing and load planning resulted in $75 million in cost savings and 72 million pounds of CO2 avoided.

    In 2020, Walmart was an INFORMS Franz Edelman Award finalist for its multiobjective price optimization framework in stores for dynamic markdown and inventory optimization.

Over a roughly four-year period, I have built – from "scratch" – a world-class team of ~15–20 talented, committed, and greatly appreciated data scientists who are focused on end-to-end fulfillment domain areas, including:

- Transportation network activity forecasting and resource utilization optimization
- Supply chain warehouse facility operations, inventory, and labor optimization
- Last mile (on-demand) delivery operations optimization
- Returns optimization
- Asset protection optimization
- Safety optimization

Due to confidentiality restrictions, I cannot share non-publicly available information, but I can say that, over the past four years, this team has designed, developed, and deployed *dozens* of data science solution applications utilizing the entire range of ML, AI, statistical, and O.R. models to deliver, at the time of publication, *over $1 billion* in supply chain-related selling, general, and administrative (SG&A) cost avoidance.

The primary lessons I have learned while working as director of data science at Walmart include the importance of:

- Available, accessible, high-integrity data.
  - Walmart is quite adept and well prepared in this regard with the large majority of supply chain data stored in a cloud platform accessible by advanced query tools.
- Business value and economic impact focus to ensure maximal productivity and output of every single data science resource and project endeavor.
- Strong partnerships and collaboration between data scientists and the wide range of stakeholders and constituents, including:
  - Business leaders and individual contributors.
  - Business-side data science teams.
  - Other data science teams working in similar, related, or tangential domains.
  - Data engineering.
  - Enterprise application teams.
    - Data science solutions are often deployed as *microservices* that are integrated with larger enterprise applications and ecosystems via data pipelines.

My fervent hope, and plan at present, is to finish my career at Walmart Global Tech in data science. There is no better place that I can presently see to be engaged in one of the largest digital transformations in history, specifically of a brick-and-mortar retailer into an omnichannel retailer, where data, data science, and analytics play such a significant and integral part in the continuous improvement of operational and economic efficiency, specifically in the supply chain domain.

Ride the Waves

In 2004, after deciding to pursue my MBA at age 40 from SMU's Cox School of Business and finishing in the top 3 of my cohort by GPA, I did have aspirations of becoming a general manager (GM) of a division or even CEO of a technology company. At the time, I was, in effect, a deputy GM (#2) (i.e., VP of engineering and product development) involved in a spinoff of a software product division of McAfee (then Network Associates), and when my GM (Jeff Honeycomb) planned to retire, I was meant to succeed him. However, instead of a "carve out" (obtain financing/operate independently), the business unit was acquired by BMC Software, and I exited a year later with a package. That turned out to be my last opportunity for that type of GM job. Without GM/COO/VP sales experience beyond startups, getting hired into that type of position is practically impossible, from my perspective, unless you start the company, work there for decades, or know the CEO and get hired into the company.

Around 2011, when the movie "Moneyball" came out, followed by "The Big Short" in 2015, I read Tom Davenport's landmark book, "Competing on Analytics" (2007 edition), and the trend toward analytics was really starting to take off. At that time, I decided to *lean in* to my O.R./statistics roots and ride the "analytics wave," much like I rode the airline O.R. wave in the 1980s and 1990s and the e-commerce wave in the 1990s and 2000s.

From 2014 to the present, I have been gainfully employed and thriving in the last phase of my career as a practitioner, leader, professor, advisor, and now two-time author sharing what I have learned about working in the analytical sciences. In 2016–2020, I developed and taught a Business Analytics elective course to EMBAs at SMU's Cox School of Business as an adjunct faculty member with considerable success, enabling most of my students to achieve significant, tangible business value and economic impact for their enterprises. I adapted that course into a Business Analytics course for SMU M.S. in data science programs. Now that the AI wave is "crashing onto the beach," I am teaching seminars on AI Strategy at SMU as part of Executive and Continuing and Professional Education programs and plan to offer an elective course on the same topic to EMBAs at SMU Cox starting in January 2025.

If nothing else, the *longevity* of my career in the analytical sciences proves just how *timeless and valuable* these lessons, skills, and capabilities truly are. Technologies and methodologies evolve, data volumes grow, computing power increases, and businesses adapt to rely more and more on data-driven, model-based problem-solving, decision-making, and question-answering, all of which create a greater synergistic, upward spiral effect on the business value and economic impact that can be created inside enterprises with data science.

Honestly, I could not be happier to finish my career working in this very highly specialized and valuable *niche* and helping to launch the next epoch of analytical sciences and help, in some small way, to prepare the next generation of practitioners, leaders, and executives looking to gain strategic competitive advantage using data science and analytics! In my case, the career journey came full circle, ending up doing what I was *destined* to do all along, albeit now as a leader, educator, advisor, and author!

2

The Dual Challenge of the Analytical Sciences Practitioner

Introduction

I wrote the following article in 1991, about four years into my career at American Airlines (AA) Decision Technologies. At this time, I realized the primary goal of an analytics practitioner should be researching (i.e., theoretical) *and* developing (i.e., applied) real-world data-driven, model-based solutions that solve problems more efficiently and effectively, help make better decisions, answer key business questions, and deliver business value and economic impact, and are *actually utilized* by their intended customer stakeholder group.

The subject matter of the article was "O.R." (operations research), a precursor to analytics – as statistics was a precursor to data science, which includes now fields like machine learning. But the *key point* remains true today, that the analytical sciences, including AI, are a *means to an end* of creating significant business value and economic impact. They are not an end themselves.

At that time, O.R. was struggling to move away from having evolved into a largely academic, theoretical discipline since its founding in the 1940s. The trend accelerated in the 1970s through 1990s with the excellent applied work done in the energy industry (oil & gas, electric), transportation (airlines & trucking), building on the work done in the 1960s in manufacturing (e.g., the seminal work in industrial production, inventory, and labor planning at PPG attributed to Holt, Modigliani, Muth & Simon).

AI went through similar dark periods, known as "AI Winters," in the mid-1970s and the late-1990s and early-2000s, when the discipline struggled with: (1) funding and interest levels outside of academia, (2) finding suitable problems that business could solve and wanted to invest in solving, and (3) lack of data and insufficient computing power to drive AI solutions.

DOI: 10.1201/9781003588344-3

Many of the principles that show up in my later writings are found in this early article, including:

- Understanding the business problem
- *Pareto principle* applied to a model that achieves 80% of the value for 20% of the cost
- Communicating results achieved to constituents and the public
- Dealing with complexity (computational) using fast, performant heuristic algorithms
- Adherence to strict timing, budgetary, and resource constraints (project management)
- Sometimes a model alone will suffice, sometimes a full-fledged system is required
- Trading analytical rigor for better implementation and solution marketing

Famed O.R. academic and practitioner, R. E. D. "Gene" Woolsey is quoted in the article. One of his quotes that comes to mind on that last point above is:

> A manager would prefer to live with a problem that they cannot solve, rather than implement a solution that they cannot understand.

Winning the hearts and minds of stakeholders and constituents and making sure that you bring them along with you on the journey is critical to a successful outcome.

How are you generating business value and economic impact with analytical sciences?

THE DUAL CHALLENGE OF THE OR PRACTITIONER

Operations Research is a discipline which has its origins in applying mathematical, scientific and computer techniques to solve *real world problems*. The very name, "Operations Research." comes from its original purpose, i.e., "research on operations" as coined by British researchers involved in the evaluation of military operations, such as submarine hunting and air defense resource positioning during WWII.

As the field evolved and became an academic and scholarly discipline, the focus on purely theoretical and rigorous mathematical concepts has to a certain extent suffocated the emphasis on applied problem solving. Michel Balinski, the renowned mathematical programmer, recently claimed that OR has "stagnated" as an academic discipline and its only remaining hope for revival would be "the infusion of empirical results." Russell Ackoff has long proclaimed "the death of OR" because of its failure to accept the challenge and complexity of real world problems and their solutions.

Having entered the OR profession in 1987 as a master's degreed practitioner with American Airlines Decision Technologies, I was disappointed upon attending two ORSA conferences (and reading the paper abstracts from several others) that the meetings seemed to be dominated by academics presenting primarily theoretical papers that were variations on the same old themes. I found even more disturbing in conversation with many participants that *practitioners* were actually looked down upon because of the supposed lack of theoretical rigor involved in their work. While I fully appreciate and understand the need for fundamental or *pure* research at the university level, I found disturbing the Society's lack of emphasis on and recognition of OR applications that involve solving "real world" problems.

Within the last few years, events like the Franz Edelman and ORSA Prize competitions, both of which were won this year by American Airlines, have just begun to recognize and promote the quality work done by practitioners. But, it has only been recently that articles with an *applied orientation* were focused upon or even included in *Operations Research,* the flagship journal of the Society.

While I recognize the outstanding work in applying OR to real world problems done by academics such as IBM Fellow Ellis Johnson (Crew Scheduling at American Airlines), Patrick Harker (Railroad Scheduling applications) and John Jarvis and Don Ratliff (Military/Logistics applications), I would like to recognize and make known the efforts, challenges and successes of the OR practitioner.

The purpose of this article is twofold: first, to briefly comment on the dual challenge (no pun intended) of research and development faced by the OR practitioner today; second. to challenge practitioners to take a more active role in the profession's. I am certain that practitioners can make a positive impact by going to conferences and presenting papers, participating in tutorials, publishing articles in the profession's journals, and promoting the discipline at the college and high school levels as an interesting and challenging career field.

In almost all cases, the OR practitioner is faced with the dual challenge of *performing research* on real, operational problems, with *soft constraints* and often *fuzzy data* and then, within strict time and budgetary constraints *developing a viable solution* to a user's or client's problem. And whether that solution is a one time decision analysis, a decision model or a full-scale, model-based decision support system, the challenge to successfully develop and implement a tool that will reduce costs, increase profits/revenues or improve operational efficiency is often times enormous.

OR practitioners' education and problem solving background allows them to bring structure, logical reasoning and formalization to an often times chaotic, unorganized, illogical and *political* industrial, corporate or governmental environment.

An OR practitioner's solution to an ill-defined and difficult real world problem takes on a variety of forms. The solution may be a mathematical, statistical or computer model which allows for the analysis and evaluation of a system in a controlled environment, or a relational database that better organizes the data associated with a decision process which is connected to a spreadsheet containing *what if...* scenario evaluation routines; or, it may be a full-fledged model-based decision support system which allows a user to organize and manage data, perform *what if...* scenario analysis and run a host of mathematical models which optimally evaluate various decisions.

More simply, the solution may be the organization of scattered thoughts, ideas, conflicting objectives and constraints into a more logical, coherent *decision framework* for a client who is *too close to the problem* to solve it objectively. All of these are synonymous with "research with relevance" - the motto of the OR practitioner who evaluates and solves difficult, *real world* problems using a variety of problem solving techniques and methodologies.

Granted, the OR practitioner may not always produce the most mathematically elegant, esoteric or even optimal solution to a client's problem, but, if the solution entails cost reduction, revenue enhancement or improving some measure of operational efficiency it is almost always accepted willingly by the client. Evaluating the success of an OR project or endeavor has long been disputed among practitioners. However. I personally believe that beyond the tangible benefits provided by the deliverable, its success can usually be measured by whether the client actually uses or abides by the recommendations produced by the tool that was developed.

In practice, whether this important criterion is achieved depends not so much on the level of mathematical rigor of the work, but more so on how well the OR practitioner implements and *markets* his solution to the client. This criterion equates, in *Gene Woolsey-speak,* to "You must know your client's business as well if not better than he does. lest he think you a fraud."

The limitation placed on practitioner's to find optimal solutions is often times governed by the "80-20 rule", i.e., the client will settle for 80% of the optimal solution at 20% of the cost to obtain the optimal solution.

Upon realizing that this was the way OR operated in the "real world" of business, where profit and loss are of paramount importance, I was reminded of a sign that hung on the wall of the Simulation and Analysis Division of Ketron Inc, where I worked as a programmer/intern while in college.

The sign read. *Operations Research is the art and science of giving bad answers to questions that would otherwise be given worse answers.* Although the initial reaction of most OR practitioners, myself included, is to reject such a statement as an extremely cynical and malicious threat to our profession and livelihood, upon further consideration and reflection it makes sense. Albeit, while this is not always the case, the statement can be interpreted as saying that it is better to have an *approximate answer* than to have none at all, especially when the *optimal answer* may not exist or be far too costly to determine.

The bottom line is that we live in an imperfect, complex and complicated world. As OR practitioners, we are charged to use our education, skills and abilities to bring order and structure to our often unorganized and chaotic world for the benefit of corporations, industry, government, and society. To research *and* develop relevant, viable solutions to real world problems using any variety of mathematical and computer methodologies in a timely and cost-effective manner truly is the *dual challenge* of the OR practitioner.

In writing this, I am accepting my own challenge to practitioners to become more active in the society and rekindle the *applied* nature and spirit of the discipline by writing more articles and presenting results of successful, as well as unsuccessful, OR applications, so that we may recognize our triumphs and learn from our failures.

Douglas Gray is a senior consultant with American Airlines Decision Technologies in Fort Worth. Texas.

"The Dual Challenge of the OR Practitioner," OR/ MS Today, October 1991 Issue.

3

Digital Transformation

Introduction

I wrote this article in 2019 while I was interviewing with Walmart because I was intrigued by the company's aggressive *digital transformation* journey from a brick-and-mortar retailer into an omnichannel retailer, expertly combining the physical presence of retail stores and wholesale clubs with the virtual world of e-commerce online ordering, and consumer pickup and delivery, on a global scale. I reflected on similar experiences that I had at American Airlines, Sabre, Travelocity, and several others, and recognized a few key patterns and trends.

Most companies (*more than* 80%) still have a long way to go to realize the full potential of digital transformation, and not surprisingly, because it is so very, very challenging on so many levels.

Analytics and, now more than ever before, AI play a huge part in digital transformation, alongside mobile and web.

The *"why"* of digital transformation is improving the customer experience and economic performance through greater efficiency. The *"how"* is based on the principles of *speed, scale, change, drive,* and *fearlessness* that propel the very best companies on the digital transformation journey.

I updated the article to reflect on my past 4+ years of intensive and impactful experience working in data science supporting supply chain and end-to-end fulfillment at Walmart Global Tech.

Where is your company on its journey of digital transformation?

Observations and Foundational Experiences

Digital transformation has clearly emerged as the "new black" of the business and technology style guide lexicon. CIOs are morphing into Chief *Digital* Officers. The emergence of the Chief *Data* Officer has accelerated to gather, govern, and mine all of the data coming from websites, mobile devices,

DOI: 10.1201/9781003588344-4

sensors, and enterprise applications. *Chief Analytics Officers*, or *Chief Data Scientists*, oversee the application of mathematics, statistics, machine learning, and AI to corporate strategy, tactics, and operations-related problem-solving, decision-making, key question answering, and all of the underlying technology infrastructure. It is no longer sufficient for technology to be *aligned with* the business, or *support* the business. Today, technology *is* the business, in many industries.

In a recent presentation, a speaker showed a slide entitled **"Mobile ate the world"** – a reference to the fact that everyone now "lives" on their phones and other mobile devices. The premise of "everything you need or want at your fingertips in an app, now," coupled with Wi-Fi everywhere and now 5G bandwidth, has accelerated the mandate for companies to transform their internal operations and customer experiences to be more inherently digital. The marketplace demands it. Every company should have a response.

Companies across industries are proclaiming, "We are not a bank, we are a technology company that lends people money" (or helps them manage their money with mobile banking and payment apps) or "We are not a car company, we are a technology company that enables human transportation" (digitization of automobile systems and driverless cars). A brave new world. Absolutely, in many respects, but not an entirely new phenomenon. The increased computing power and easier access to technology platforms, in particular, open source, combined with the pace and pervasiveness of new uses of new digital technologies, and the volume, velocity, and variability of data are all certainly new. However, many of the world's best companies have been using technology to digitally transform their businesses and industries for decades.

As someone who has spent their entire 35-year career making a living at the intersection of business and technology, I both applaud and smile a bit wryly at the *new age of digital transformation*. Digitizing the analog world, creating "digital twins," and harnessing the power of consumer behavioral and enterprise data offers great promise to streamline, automate, and optimize business processes, and wring out the inefficiencies that creep into companies' manual methods over time. Inefficiencies adversely impact profitability, slow down execution, and impede growth. Mobile devices and apps accessing high bandwidth networks provide companies in every industry with new dimensions of customer and employee engagement and understanding. Reflecting on the trend of digital transformation, both from the key learnings of my own experiences and those of most digital companies, I found myself asking *Why?* and *How?* do companies reinvent themselves digitally. Behaviorally, what makes these companies "tick"?

I consider myself fortunate to have worked for multiple companies that were digitally transformative at various stages throughout their evolution. American Airlines (AA) transformed travel distribution in the 1970s by providing computer terminals to travel agents that enabled them to access AA's Sabre reservation system via a private network and sell airline seat inventories. AA harnessed all of that sales data in the 1980s to transform the deregulated airline industry using analytics, then known as operations research ("OR"), to better forecast passenger demand and optimize fare prices and seat inventories to maximize revenue, known today as *Revenue Management*, and applied in hotels, cruise lines, rental cars, and even self-storage facilities. Revenue management, coupled with AA's discounted ***SuperSAAver*** fares, was so powerful that it literally put low-fare People's Express *out of business*, according to CEO Donald Burr. In the mid-1990s, after being spun off from AA, Sabre reinvented itself yet again as an early entrant into the e-commerce era with the launch of ***Travelocity***, the world's first real-time, online travel booking engine. (It seems hilariously incomprehensible now, but I can remember people actually saying in 1995, "The Internet is just another fad" and "Who would ever want to buy an airline ticket over the Internet!? Just call a travel agent!")

During the years I worked for AA/Sabre/Travelocity, we used the technologies available to us at the time, plus a lot of mathematics, to digitally transform the businesses we were operating, in the wake of the competitive anarchy sparked by airline deregulation, and the disruptive power of the internet. We streamlined, automated, and optimized spare part inventories, flight and crew schedules, aircraft maintenance and hangar schedules, and even the number of catering carts. *We even reduced jet fuel costs by optimizing the number of brands and amount of liquor we carried onboard to reduce aircraft weight!* We enabled consumers to shop for and buy travel from home.

For me, those 10 years instilled a fundamentally different way of thinking about what we now call digital transformation. Some refer to it as the *"Art of the Possible."* Intel CEO Andy Grove said it plainly, with regards to competition, that, *"If it can be done, then it will be done, and if not by you, then by your competitor(s)."* Simply put, we addressed head on how can we aggressively leverage technology to do things differently, better, faster, and cheaper, to survive, thrive, and then grow, more profitably.

The most recent 4+ years of my career have been spent at Walmart Global Tech, which continues to support Walmart's digital transformation from a brick-and-mortar retailer to a digitally enabled omnichannel retailer that serves the customer however and whenever they choose to engage: shopping in-store or in-club, picking up items at the store or club, or having groceries and general merchandise delivered to their home (and, if they choose, even put away). From robots palletizing products in distribution centers and checking inventory in stores and clubs, to driverless vehicles delivering groceries to customers, to data science being utilized to optimize the customer shopping experience online, and optimizing inventory and operations in the end-to-end fulfillment supply chain, Walmart relies *heavily* on data, technology, and analytics to continuously increase profitability through improvements in operational performance and economic efficiency.

Both AA (1991) and Walmart (2023) are winners of the ***INFORMS Prize*** and ***INFORMS Franz Edelman Award*** (as well as one or more runners-up), which is a testament to their commitment and performance in accelerating digital transformation with data and analytics.

The Internet, Data, and Analytics in Digital Transformation

Based on my experience at AA/Sabre/Travelocity of transforming businesses using technology to create business value, economic impact, and sustainable competitive advantage, I sought out and participated in a series of digital transformation endeavors over the next 15 years.

In the late 1990s, I was the CTO for one of the first pure-play e-commerce boutique consulting companies that helped to pioneer online B2B/B2C retailing of, among other goods, computer equipment. Along with companies like Computers4sure (sold to Office Depot in 2001), we took orders from businesses for PCs and servers on a website and then drop-shipped the equipment directly to the customers from distributors, cutting out middlemen, and reducing supply chain friction. *We never even touched the merchandise – a 100% digital business!*

During the dot-com era, I was the CTO for one of the first online mortgage companies that integrated a toll-free brick-and-mortar call center of mortgage specialists, with a website storefront, and an AI-based ***Personal Mortgage Optimizer*** to help a consumer find the *best* mortgage among millions of products. According to the *Rule of Three* (Business & Economics), we finished *fourth,* and our assets were bought by LendingTree. More importantly, the *industry* was digitally transforming how people shop for and buy mortgages. Rocket Mortgage has taken that idea to the next level with a quick online process for mortgage qualification and approval.

Throughout the first decade of the 2000s, my teams and I worked on a variety of game-changing apps.

- An app enabling doctors to view their patients' digital radiology scans (CT, MRI, X-ray) via an internet browser; that company evolved into the #1 DICOM image cloud storage company.
- An app for experientially planning and shopping for your next dream vacation on a mobile tablet, harnessing NLP, internet search, and AI to create the ideal vacation package; another company, Wayblazer productized this concept, using IBM's Watson platform.
- An app for a $2 billion healthcare insurance company, and now hospitals, using patient medical and demographic history data to predict the likelihood that the patient will need to be readmitted to the hospital after surgery or drug rehab, require spinal surgery at some point in the future, or purchase a new dental insurance product.

According to the *McKinsey Global Institute's Industry Digitization Index,* the U.S. is currently only operating at 18% of its digital transformation potential. Not surprisingly, because many industries have a long way to go to digitally transform their operations and customer experiences. Yet, therein lies the opportunity for digital technologies to transform business models. Think IoT – sensors, video cameras, mobile devices. Think robotics, robotic process automation, and AI, machine learning and analytics – combining vast amounts of data, practically unlimited bandwidth, and massively parallel processing (MPP) and GPU (see NVIDIA) computing horsepower, and some of the most powerful mathematics and computer science models and algorithms ever devised embedded in enterprise systems to streamline, automate, and optimize a company's most mission-critical, high value-add processes.

Unfortunately, only 20% of the data science and analytics models that get built today actually get implemented into a "production system." Why is that? Several reasons, including (1) model builders are not always closely aligned with the IT Department that builds enterprise systems, (2) often the enterprise data infrastructure is not sufficiently mature to support ongoing model utilization, (3) building analytical model-based enterprise systems, for planning applications, let alone real-time applications (think stock trading systems), is more than 10–100x as complex, time-consuming, and costly than just building the model alone! Tom Davenport, author of *Competing on Analytics*, said "Models make the enterprise *smarter*, but models embedded in business processes and systems make the enterprise more *economically efficient.*" It is this type of economic efficiency, i.e., producing a greater number of units of output of higher quality in exchange for a fewer number of units input(s), faster, that is one of the key goals of digital transformation, along with digitally bonding your firm to your customer to build a relationship.

You can trust me on this point, having spent 10 years leading teams building enterprise-scale planning and real-time analytics model-based systems for U.S. airlines. In one case, it took over eight years to gather all of the data, integrate all of the systems, and build and test the models and algorithms to deploy the world's first real-time airline network irregular operations recovery optimization system. The *good news* is that we significantly increased on-time performance, especially markedly during major winter storms, reduced the cost of weather and other unanticipated network disruptions, improved the customer experience by reducing flight delays and cancellations, reduced passenger itinerary disruptions, and provided 24-hour advanced notification and rebooking in the wake of cancellations, and increased aircrew member quality of life. We had similar successes with analytics model-based systems that reduced crew and jet fuel costs, the two largest (operating expense) cost categories at every airline.

At Walmart, the success rate of data science projects is considerably higher for several reasons, including a closely integrated working relationship between data scientists, business stakeholders, and data and technology constituents to understand the business problem at hand and agree on the appropriate modeling approach, a focus on and prioritization of projects based on business value and economic impact upfront before projects begin and tracking value captured through project completion, and robust enterprise technology platforms for deploying models and systems to production where the value is realized. These attributes are embodied from senior leadership all the way through to the engineers who do the heavy lifting to manage extraordinary volumes of data, build models and software, and deploy and operate end-user facing solutions.

The Why and How of Digital Transformation

Why do companies choose to invest and digitally transform themselves?

Two motivations are clear.

First, to provide a dramatically improved customer experience to better satisfy the consumer's insatiable demand for immediate gratification in practically all things, i.e., service/product delivery, the need to know *now*. This mandates information, and data, which is either the product itself, e.g., music/movie streaming or a banking transaction, or *about* the product, e.g., shipment status-ETA, must be able to flow as 1s and 0s at lightspeed to the customer's mobile app. Increased levels of more holistic customer engagement lead to greater customer satisfaction and loyalty, which results in more purchases and greater customer lifetime value. The *virtuous cycle* (customer positive reinforcement) is completed by the mobile app.

Second, to wring out the inefficiencies of (manual, spreadsheet-based, or "batch" system) internal business processes that create friction, slow down value-added operations with non-value-added steps, and adversely affect not only profitability, but a company's inherent ability to be nimble, proactive, responsive, and compete effectively and efficiently. ERP, SFA, CRM, etc., were intended to solve this problem but fell short. Robotic Process Automation (RPA) is the latest technology to address "swivel chair interfaces," which can and should augment continuous process improvement. Predictive and prescriptive analytics models can automate the anticipation of what is going to happen next, and optimization of the outcome, e.g., minimizing the time and cost of shipping a product to the customer.

How do they do it? Behaviorally speaking, what traits do those companies have in common?

Digitally transformative companies share traits that are inherent in their people and/or culturally mandated by aggressive, visionary leaders, e.g., Elon Musk at Tesla and SpaceX.

1. *Speed*. A leading technology company that loans people money, i.e., a "digitally transformed bank," indicated that its move of 100% of its technology infrastructure and Agile/DevOps-based processes to the cloud was primarily about *speed*. Like the real estate mantra,

"Location. Location. Location.," this company's top three decision criteria are "Speed. Speed. Speed." Speed of execution, speed of time to market new products and services, speed of value delivery internally and to the customer, speed in the customer's experience... you get the idea. Move fast, change and adapt quickly, operate at the pace of the digital customer, and faster than the competition is the goal. *Speed* can only be achieved if you're *digital.*

2. *Scale.* When your business is digitally enabled with platforms for product and service delivery that scale without the addition of capital expenditures or vast amounts of labor, growth can be realized much more rapidly, and much more profitably. Profitable growth is every CEO's primary goal. When Cisco was going through a period of particularly heady and rapid growth in demand for routers and switches, the CFO calculated that they would need *33,000 additional engineers* to configure its products to meet customers' requirements. Realizing that was infeasible, on multiple levels, Cisco built a website, backed by a set of AI-based algorithms, that could automatically configure networking gear products to a customer's specification *without human intervention.* Profitable growth and business model scalability were achieved by digitally transforming a manual, human-intensive, and complex business process.
3. *Change.* Digitally transformative companies are comfortable with continuous change and adaptation, at scale, and at their chosen speed of execution. Think Darwin – Survival of the Fittest through Adaptation. Think Schumpeter – Creative Destruction and self-imposed disruption of your own business model. Think Grove – Only the Paranoid Survive. Think Google from online search to driverless vehicles. These companies *thrive on change* as a means to an end of continuously improving performance in terms of customer experience and profitability. (The Japanese call this *kaizen.*) Changes such as in business processes and practices, people and culture, and technology platforms and approaches. Digitally transformative companies will implement any change that is conceivable and warranted to more efficiently and effectively achieve speed and profitable growth. If they can't buy it, then they invent it themselves. (Alternatively, Walmart bought Jet.com to accelerate the company's transformation to e-commerce and omnichannel retail, while leveraging its physical stores for distribution and pickup operations; *90% of the U.S. population lives within 10 miles of a Walmart,* which provides incredible customer reach and high touch augmented by online offerings.)
4. *Drive.* Driven to compete and win every day. From Bob Crandall at American Airlines and Sabre to Elon Musk at Tesla, companies that practice digital transformation are driven to *win.* Pure and simple. Win their customers, win market share, win for their employees, win in the financial/capital markets. They do not shrink from making big,

bold investments in all available digital technologies that are relevant to their business. They are not afraid to invent, reinvent, and try new things, and fail. Quite the opposite, they fail fast and learn quickly from their mistakes, they continuously tune and move forward, or abandon an approach that has been proven not to work, and try something new. This kind of drive comes with a rare balance of will and grace, extraordinary confidence and equal humility, and both IQ & EQ.

5. *Fearlessness.* Digitally transformative companies are inherently more comfortable, and frankly fearless, when it comes to leveraging all sorts of technology platforms, and technology-based processes and ways of doing business. Even more importantly, they are comfortable continually rethinking and remaking their entire end-to-end analog, or old-school digital, business model, processes, and delivery execution leveraging the most impactful new digital technologies. This digitally focused approach is an absolute imperative for faster customer value delivery and internal business value creation, more profitable growth by avoiding friction, drag and inefficiency with mission-critical processes that are streamlined, automated, and optimized. When your business is digital, i.e., *virtual*, based on *"code,"* the ability to change, adapt, respond and, if necessary, react is an order of magnitude more easily attained.

Some will say that digital transformation is easier for "service businesses" such as banks and airlines, than it is for manufacturing companies. Toyota pioneered, and is perfecting, robotics in automobile manufacturing, enabling Toyota, and luxury car division Lexus, to regularly outperform all of their competitors in sales volume, quality, customer satisfaction, and profitability, i.e., economic efficiency – producing a greater number of units of output of higher quality in exchange for a fewer number of units input(s), faster. I had a student in an EMBA Business Analytics course that I teach at SMU who was a steel mill deputy superintendent. The mill was struggling with profitability. Using customer product demand data, mill production data, and mill product profitability data, the student applied mathematical programming to optimize the mill product mix and production schedules. The mill increased profitability by 23% and the deputy superintendent was promoted and rolled out his solution to the company's other steel mills! *If it can be done, it will be done!*

Digital Transformation: A Competitive Imperative

More companies than ever are now embarking on digital transformation initiatives due to customer demands, competitive pressures, and economic necessity. And, of course, the number and range of digital technologies that are available today, and their myriad uses and impacts, are exploding at a rate of change that is faster than ever. As important as the technologies are in digital transformation, the point is not as much about the technologies as it is about the *focus* and *behaviors* in employing the technologies. The companies that digitally transformed themselves and their industries over the past five decades had the same timeless and fundamental traits as Toyota: a penchant for speed, scale, change, drive, and fearlessness in transforming from analog to digital to achieve greater economic efficiency and bond themselves in a more engaging relationship with their customers, and partners.

Anyone engaged in digital transformation knows how incredibly difficult it is to do at all, let alone do well. The amount of change and disruption is *cataclysmic*, the speed is *breakneck*, and the risk is *high*, if not properly managed. The *complexity* involved and the sophistication required for success are equally high and can seem daunting. The Navy SEALs, America's most elite fighting force, have mantras for dealing with this kind of pain:

- Embrace the suck
- Get comfortable being uncomfortable
- The only easy day was yesterday

Sound advice for those digitally transforming an enterprise or an industry.

Notwithstanding the tremendous investment in time, resources, and effort, and the pain endured in executing a digital transformation, the associated benefits are significant and substantial. You are better preparing your company to engage more adeptly with digital customers, employees, and partners. You are enabling your company to execute and deliver business and commercial value to customers faster, more consistently, and be more proactive and responsive. By streamlining, automating, and optimizing internal workflows, you are setting a precedent for a more scalable business model that can lead to reduced costs, increased profitability, and increased attainable growth. By deepening multichannel information flows and intensifying engagement via high value-add mobile apps and devices, you are building longer-lasting, more mutually beneficial relationships with customers, employees, and partners.

Digital transformation is indeed a strategic enabler, but not a panacea. No company's success can be guaranteed, even by something so pervasively and favorably impactful. Digital transformation is a competitive imperative. At a minimum, your enterprise realizes a dramatically increased chance of surviving and remaining economically viable and competitively relevant in the decades to come. At a maximum, your company can become economically and competitively more dominant, even predominant, in your industry, by leveraging the speed and value multiplier afforded by the power of being digital. The power to leapfrog your incumbent competitors, and stay ahead of innovative, unconstrained startups seeking to disrupt you out of existence is attainable. Reaching a point anywhere along that spectrum is a win.

Consider three examples: **American Airlines** not only survived the deregulation cataclysm of the 1980s by digitally transforming themselves with automation, data, and analytics, when so many other airlines did not (TWA, Eastern, Braniff, Pan Am, and PEOPLExpress to name a few), but went on to thrive and become one of the largest, most dominant air carriers in the industry. (Surveys show Delta Airlines has caught up and surpassed them since – they implemented their own versions of revenue management and analytics in parallel to AA.) **Walmart** could not even feasibly operate their global omnichannel retail business at the scale they do today, let alone do so profitably, had they not maximally leveraged web, software, data, analytics, and cloud data center technologies to digitally enable their customer shopping, buying, and fulfillment experience, and optimize their end-to-end supply chain processes. Lastly, **Toyota** has become the dominant global automobile manufacturer, by most metrics, including total sales, quality, and customer satisfaction. Robotics, AI, and heavy automation, along with *Kanban* just-in-time manufacturing and *kaizen* continuous improvement methods, played an enormous role in that accomplishment over several decades.

There is one hope for us all. Digital transformation does not have to happen all at once. No one wakes up one day "digitally transformed." Like most things in life, e.g., hitting a baseball, learning to ice skate, learning to code, digital transformation is a *journey, process, mindset,* and *a way of thinking and behaving.* It takes a significant amount of time and effort to get started, get moving, really get rolling delivering value and hitting milestones, and ultimately be successful, then maintain at that level, and then get to the next level. Will, tenacity, and resiliency are more important than intellect, although serious technology skills are required.

Start now. Or not. The choice is yours. Just remember this…

"Only 53 companies have been on the Fortune 500 since 1955, thanks to the creative destruction that fuels economic prosperity." – Mark Perry, American Enterprise Institute

4

Advanced Analytics is Economically Transformational

Introduction

Continuing on the theme of *transformation,* I wrote this article as a motivator for my students and colleagues to expound on the *"why"* behind analytics, which is that it is *economically transformational.* Companies and entire industries have been transformed by advanced analytics. The best example I witnessed was yield (revenue) management in the airline industry pioneered at American Airlines (and Delta contemporaneously) in the 1980s, which transformed dynamic, competitive pricing and commodity seat inventory management to maximize revenue on every flight. That methodology is used today in hotels, cruise lines, rental car companies, tour companies, passenger and freight railroads, movie theaters, and even self-storage facilities!

What economic performance metrics can advanced analytics measurably transform in your business and industry?

What if I told you that your company could...

- *Increase sales closure rates by 30% and double market share? (Targeting the right prospects)*
- *Reduce transportation costs by $6 million? (Optimizing truck delivery assignments and routes)*
- *Predict hospital post-surgical readmissions with 90%+ accuracy, reducing readmissions-related costs by $500,000 annually, while improving patient outcome and quality of care?*
- *Increase annual revenue 4%–6% annually on $14 billion in revenue? (More accurately forecasting demand, dynamically adjusting pricing and optimizing inventory allocation)*
- *Reduce operating costs $100 million annually on $20 billion in revenue? (Optimizing high-skilled labor utilization & supply chain operations)*

DOI: 10.1201/9781003588344-5

If you are like most people, you would probably smile politely in disbelief, and say, "No way!" An expected and reasonable response. These types of results seem practically unbelievable, right? (BTW, they're *all* real.) If you're inquisitive, and inclined to want to achieve similar results, you might ask, "How?"

The answer is… analytics. Specifically, *advanced* analytics, i.e., *predictive* analytics – answering the question, *"What outcome is going to happen in the future with what odds or probability?"* –and *prescriptive* analytics – answering the question, *"How do we optimize the outcome of what is going to happen?"*

Advanced analytics, i.e., mathematics, statistics, computer science models, and algorithms, coupled with software and computer technology, and LOTS of data about your customers and your business operations, is one of the most economically impactful elements of digital transformation available. *Economic efficiency,* i.e., producing a greater number of units of output of higher quality in exchange for a fewer number of units input(s), faster, is one of the goals of digital transformation. *Advanced analytics delivers on this goal. IDC reported that the median ROI of BI-only projects is 89%; incorporating analytics raises the median ROI to 145%.*

How? The *$64 million question,* literally! First, predictive analytics.

Baseball Hall of Fame catcher Yogi Berra, famous for his "Yogi-isms," said, *"It's tough to make predictions, especially about the future."* (Nobel Prize-winning Physicist Niels Bohr said something similar, but Yogi was more entertaining!) Business, like life, is full of uncertainty. That uncertainty stems from complexity. No human alone can consistently and accurately predict what is going to happen tomorrow, next week, let alone next year. Think weather, oil prices, demand for a (new) product, etc. (If you could, you'd clean up on Wall Street and the lottery!) There are too many factors, and too many variables to consider, and often those variables change over time, and interact with and depend on each other.

However, there is an approach to collect and analyze as much data about as many factors and variables from the past as possible to build *models* that can identify trends, patterns, and predict the *odds* or *probability* of certain outcomes. The *goal* of predictive analytics is to get a lot better than "flipping a coin" (50-50) to predict an outcome, say 80%–90% accurately, knowing full well that your predictions will never be 100% accurate. Pragmatically, models will need to adapt, evolve, and adjust to changing conditions over time. George E. P. Box, statistician and Father of Time Series Analysis (Forecasting), famously said, *"All models are wrong, but some are useful."* (Every model has some predictive "error," but some are *far* better than flipping a coin.)

- Out of 100 patients, if you can predict with 90% accuracy the 30% that have the highest probability of readmitting after surgery, you can focus your limited patient care staff on those patients, and pay less attention to the other patients that are less likely to have issues.

- Imagine being able to predict with 90% accuracy the on-time performance of your commercial airline tomorrow at every airport as a function of weather and passenger traffic load; decision-making concerning the recovery of trouble areas would be far more effective.

In both instances, you will get it wrong 10% of the time, but that is a lot better than 50% by guessing!

Second, prescriptive analytics.

Businesses have *objectives*, e.g., maximize revenue, minimize costs, and *constraints*, e.g., limits on resources, like raw materials, staff, and productive capacity (steel mill, airplanes) or time windows when events must occur, e.g., deliveries. *Variables*, such as how much of each product to make and sell at what price, are difficult to determine visually or in a spreadsheet to reach the optimal outcome due to the vast number of possible combinations and permutations of variables and their values, that constitute the range of solution outcomes. The field of mathematical optimization, known as *mathematical programming*, was developed to solve such problems.

One incredibly insightful real-world application of mathematical programming, and its impact on business value and economic efficiency, is the one below from United Parcel Service (UPS). Every day, UPS must optimize the routes of *55,000 delivery truck drivers* – no small feat solving that problem!

For UPS, eliminating one mile, per driver, per day over one year can save up to **$50 million**. By the end of 2016, 55,000 routes optimized daily by the On-Road Integrated Optimization and Navigation (ORION) system will have saved 10 million gallons of fuel annually, reduced 100,000 metric tons in CO_2 emissions, and avoided an estimated $300 million to $400 million in costs.

Often times in analyzing a complex system with many moving parts that interact – and events and activities are often random or probabilistic in their occurrence and duration – it is not possible to formulate a mathematical optimization problem with equations representing objectives, variables, and constraints. In those instances, we utilize other prescriptive analytical techniques, such as discrete-event (aka, *Monte Carlo*) computer simulation to optimize the policies that govern the operation of the complex system, e.g., cranes unloading ships in a harbor cargo dock, customers waiting in line at a bank or amusement park, or operating an airline schedule or an airport. By simulating the system's operation under different conditions with thousands of computer simulation replications, statistically, we can measure and determine which policies work *best* to achieve the desired business outcomes.

Advanced analytics is economically transformational. Advanced analytics, coupled with large amounts of data and readily available inexpensive computing power, possesses a uniquely powerful capability to cut through the uncertainty and complexity that grip business operations, clarify the likelihood of outcomes, prescribe the best actions to take, and deliver significant, tangible, measurable business value and economic impact. This thesis has been proven consistently over decades in industries ranging from transportation (especially airlines), retailing, telecommunications, CPG, advertising, manufacturing, financial services, and healthcare.

The examples provided at the beginning of the article are *actual real-world outcomes* either from companies I have worked for myself, or SMU EMBA Business Analytics students have worked for, and completed as projects in the course I teach annually.

Advanced analytics is now more popular, and more accessible, than ever given the vast amount of data and computing power available, to accompany both classic and new models, algorithms, and technologies. Every industry, and every department in every company, can benefit significantly from advanced analytics: Operations, Marketing, HR, Finance, Supply Chain, etc.

However, engaging in advanced analytics is not for those lacking will, nor are the results achieved without years of investment and dedicated effort by highly-skilled professionals. Speaking as someone who has made the journey in multiple industries, the benefits achieved were worth the effort.

For more on analytics, see the series of best-selling books by world-renowned researcher and best-selling author Tom Davenport, Ph.D., **Competing on Analytics, Analytics at Work, Keeping Up with the Quants, The AI Advantage, Working with AI,** *and* **All in on AI**. *And for analytics in popular culture, see the books, and movies, by best-selling author Michael Lewis,* **Moneyball** *and* **The Big Short**.

5

Airworthy: American Airlines Heavy Maintenance Planning and Scheduling

Introduction

There are defining moments in every person's career (some favorable, others not). This article summarizes one of my first favorable defining moments. A project that: (1) got me recognized as someone that could work closely and effectively with customers, solve a business problem that was causing considerable pain, deliver a holistic, stand-alone pain-relieving solution, and generate significant business value, (2) led to my first big promotion to Manager/Principal at AADT, and (3) put me on a trajectory to senior leadership.

I published this article in 1992 to commit to paper my first successful model-based solution design, development, and deployment from "scratch." This was the very first time I coded an *entire system* starting with a blank screen C compiler/development environment. As daunting as that was, the experience was *invaluable* and *foundational* to my career as a practitioner and leader, influencing how to go about getting things done that leads to a favorable outcome for all involved.

The story of the project, the approach, and outcome is detailed in the article, and I alluded to it in Chapter 1, so I won't give away the ending here.

What I will say, to characterize the story from a career development lessons learned perspective, is that while many of my peers and colleagues were fortunate to work in the far more glamorous, some might say "sexier," areas of airline O.R., like yield management, flight scheduling, even crew scheduling, I ended up on the *operations* side of the business working in, some might say, the "grungier backwaters" of airline O.R., like airport operations, airline operations, crew training, and aircraft maintenance. First of all, it is important to note that if these "operations" domains and disciplines are not running smoothly and empowered by analytical, automated solutions, all of the "sexier" O.R. quickly becomes *impaired*, i.e., metaphorically speaking, *goes right into the dumpster where it catches on fire.* Airports and airlines have to operate in bad weather, pilots need to be trained, and planes need to be maintained, or no one is going anywhere. Secondly, the moral of this story

DOI: 10.1201/9781003588344-6

is that you, as a practitioner with help from a capable software engineer, can create *significant value* with analytics working in the "grungier, less glamorous parts" of your company, to the tune of *over $1 billion* (2023 dollars), in this case, over the life of an airline fleet, with a very fast, heuristic algorithm solving a classic problem model formulation, and a very visually appealing and functionally efficient GUI-based Apple Macintosh PC software application.

Look for opportunities with *large value multipliers* wherever you happen to land in your company, preferably where people are currently solving complex problems with *big sheets of paper and colored pencils or spreadsheets*, and then … make the most of it.

Read on and you will see exactly what I am talking about. *And, by the way, the O.R. and systems approach employed was quite clever, if I do say so myself – credit Georgia Tech ISyE Drs. Jarvis and Ratliff for their inspiration in the field of interactive optimization* (circa 1986).

AIRWORTHY: DECISION SUPPORT FOR AIRCRAFT OVERHAUL MAINTENANCE PLANNING

American Airlines' fleet of approximately 600 aircraft consists of 10 different fleet types including Boeing 727, 737, 747, 757, 767, and McDonnell Douglas (MD) Super 80, DC-10, MD-11, Airbus 300, and Fokker 100. The 727 and 737 are in the process of being retired and replaced by the quieter, more fuel efficient Super 80 and Fokker 100 aircraft. The DC-10 slated for retirement in the late 1990s will be replaced by the larger, longer range MD-11 aircraft. All aircraft are at various stages of existence, with each fleet having its own unique utilization and maintenance profile.

Maintenance requirements for intensively utilized, commercial aircraft are extensive. Over 30 different types of maintenance checks are required periodically to continually ensure the airworthiness of each aircraft in the fleet. The most costly of these checks is the overhaul check, a.k.a., the heavy and light C checks or main base visits (MBV), in which the entire aircraft is essentially rebuilt from scratch. Overhaul checks of this type are extremely expensive, costing AA several hundred thousand dollars for MBV and up to $1 million for heavy C overhaul checks. The magnitude of these costs is driven by the type and age of the aircraft fleet. The frequency of these checks ranges from one MBV every 18 months for a DC-10 to one heavy C every five years for a 727, with 727 light C overhaul checks occurring once every year.

AA's Maintenance and Engineering (M&E) Long Range Planning (LRP) group develops and maintains a 5-year planning horizon schedule of base maintenance activity. This schedule, which tracks when aircraft maintenance will be performed, includes aircraft overhauls, primary component (e.g., landing gears, flaps) removals and special visits, all of which are performed at the main maintenance bases in Tulsa, Okla., and Alliance Airport in Fort Worth, Texas. This 5-year plan, also known as the "dock plan," is in a state of flux as it continuously evolves due to the fleet's constantly changing size (due to retirements and new deliveries), composition (various fleet types), utilization (seasonal changes) and maintenance requirements.

Overhaul maintenance planning at AA M&E LRP was historically a manual process. The rapid growth of the AA fleet and its corresponding complexities outpaced the evolution of the dock plan development process. After paper-and-pencil methods were deemed no longer feasible, fleet maintenance and utilization data were imported into a computerized spreadsheet and operated upon to generate relatively low-quality dock plans. Several iterations, manual overrides and adjustments were necessary just to generate an inaccurate plan that was often out of date before it could even be distributed.

American Airlines Decision Technologies (AADT) was requested by M&E LRP to formalize, automate and enhance the maintenance planning process through the development and implementation of a decision support system to aid maintenance planners in the generation of the dock plan. Project guidelines stipulated a 1-man-year manpower constraint, a 6-elapsed-month development time frame, and the system could require no new hardware or software due to budgetary constraints.

AIRCRAFT MAINTENANCE PLANNING

Maintenance programs, including overhauls and component removals, are performed periodically according to time measurement specifications. For example, a 727 must undergo a heavy C check overhaul every 14,000 flight hours and DC-10 landing gears must be removed every 7,900 cycles. Occurrences of these maintenance programs can be forecast using estimates of daily aircraft utilization. Knowing the daily utilization of an aircraft and the number of hours or cycles allowed between overhaul checks, the date on which the aircraft will reach its allowable limit can be readily computed.

Maintenance Check Yield is the primary measure of effectiveness of a maintenance plan used by dock planners. Maintenance Check Yield is defined as the number of hours or cycles that an aircraft has flown between consecutive maintenance base visits for a particular type of maintenance program. Yield may also be interpreted as the percentage of allowable hours or cycles that an aircraft has flown when it is presented at the maintenance base for a certain program, e.g., 727 tail number 123 has 2,500 hours towards light C check No. 3 when it is allowed up to 3,000 hours, which results in a yield of 83 percent. Imposed upper and lower yield control limits determine when to schedule aircraft for maintenance in order to optimize maintenance program yields.

Maintenance Facility Capacity to perform aircraft maintenance is limited in quantity (e.g., number of hangars, mechanics, equipment), capability (e.g., fleet type, program type limitations) and availability, as well as expensive to support and maintain.

Therefore, the objective of aircraft maintenance planning is to optimize maintenance program yields (i.e., maximize yields between their respective lower and upper allowable control limits) with a minimal amount of maintenance facility capacity subject to a variety of operational constraints. These constraints include: similar fleet types must be processed contiguously in maintenance facilities, similar maintenance programs must be performed contiguously in maintenance facilities, and facility utilization must be constant, with little or no downtime.

Typical fleet maintenance planning includes a number of "What-if" scenarios:

- a fleet is grounded (i.e., daily utilization of 0.0 hours);
- seasonal changes alter aircraft utilization;
- the number of retirements changes drastically over the next five years;
- the number of new deliveries changes drastically over the next five years;
- the construction of a hangar is delayed;
- a new semi-annual FAA structural check is issued for the 727 fleet;
- 727-023 heavy C checks begin to require five weeks instead of the usual four;
- allowable limits increase/decrease on a particular maintenance program.

DockPlan System Development

The first step in developing DockPlan was to organize the process by which a maintenance plan is developed. Second came the organization of the data associated with generating a schedule of overhauls into a coherent, manageable framework of profiles, e.g., Fleet Utilization Profile and Fleet Maintenance Profile. Next came the identification of the goals and objectives of what the schedule should be as well as the constraints which hinder the achievement of the objectives.

With these three elements, we set out to design and develop a flexible, user-friendly decision support system which would allow the user to generate a maintenance plan and then perform various what-if analyses to evaluate the impact on the plan of changes in any of the key maintenance variables. The need for flexibility was of paramount importance due to the need for planners to react quickly and deftly to the rapidly changing maintenance planning environment. The system would be required to have three levels of functionality, including; 1. manage and develop maintenance planning data, 2. generate maintenance plan scenarios quickly and efficiently, and 3. generate a variety of tabular and graphical reports to describe and evaluate the maintenance plan.

DockPlan System Overview

At the request of the user, the entire system was designed for and implemented on existing software and hardware. Fortunately, M&E LRP already had a sufficient Apple Macintosh IIcx computing platform in place on which to build the system.

Using standard C++ and Macintosh features, an object-oriented, menu-driven, windows-based user interface was created to drive the system and allow for maintenance data table development. The scheduling model and report generator portions were developed in Think C, an ANSI Standard Version of the C programming language intended especially for the Macintosh. Excel spreadsheets, with which the primary users were already intimately familiar, were used as a format for all of the tabular output reports to allow for any additional ad hoc output analyses.

DockPlan consists of four modules. The system includes a variety of features to enhance the dock plan development and evaluation process:

- flexible to handle various problem sizes (fleet size, number of maintenance facilities and number of maintenance programs), limited only by computer memory capabilities;
- monthly specification of daily aircraft utilization rates;
- allows for new maintenance programs to be added and scheduled;
- variable yield control limit specification;
- reconfiguration, retirement and new delivery programs are considered;
- adjustment of quantity of maintenance facilities available to compen5ate for rise and fall of maintenance demand; and
- input data checking and user warnings of errors or discrepancies.

Scheduling Model Methodology

The conceptual methodology of the DockPlan system was designed to combine a planner's knowledge of and experience in overhaul scheduling with the computational power of a simple, computerized scheduling heuristic algorithm. This combination was pursued to provide the capability to quickly generate 5-year plan overhaul maintenance schedules for user evaluation. This methodology, which relies on a human's evaluative and judgmental capabilities, and a computer's number-crunching power, represents an innovative approach to the solution of complex, real-world scheduling problems complicated by a large number of dynamic, interactive variables and soft constraints.

This methodology, known as interactive optimization, attempts to reach an optimal or near-optimal result (in this case a high-quality overhaul schedule) through an iterative, interactive process linking an experienced planner to a powerful, computational schedule generation tool. Such an approach is extremely effective when user-controlled availability of resources helps to drive the solution "in the right direction" and solution quality can be evaluated by inspection using a graphical user interface.

Optimization methodology provides for the formulation of a problem into objectives to be striven toward, subject to operational and resource constraints. However, it is often the case that the size of the problem and its variable interaction complexities complicate the use of a formal, mathematical programming formulation and solution approach. Although the objective of the scheduling problem is to optimize check yields subject to operational constraints, the process of achieving this objective is confounded by the number of checks to be performed at any given time and the number of hangar spaces that are made available at that time. Therefore, a hybrid modeling approach which combines the normative techniques of optimization-based heuristic algorithms with the evaluative methods of simulation is more appropriate.

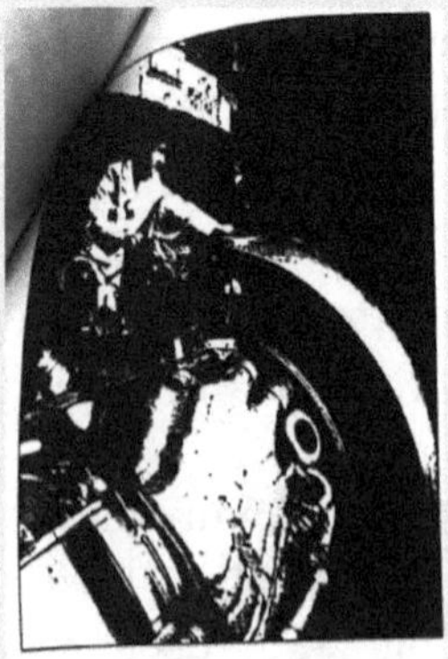

A heavy C check can cost up to $1 million.

In this methodology, the user initially specifies the parameters of the problem (e.g., such as how many hangar spaces will be available throughout the planning horizon) and then applies the model's simple, heuristic algorithm to quickly generate a schedule. The heuristic attempts to optimize the yield of each check as best it can, one at a time, as they fall due in the planning horizon, given the capacity initially made available by the user. The user then reviews the dock plan solution, adjusts the parameters accordingly, and re-runs the model to generate a new, and hopefully improved, solution.

The planner continues to iteratively run the model, adjusting parameters so as to "push" the schedule generation process in the right direction in order to reach an optimal or near optimal solution; i.e., a schedule in which most checks have yields near 100 percent and no checks exceed their allowable limit. This interactive approach also augments the capability to perform sensitivity analysis on model parameters which may change often during the course of the planning horizon.

This interactive approach has one not so obvious, but important, benefit. In the first stage of model development, it is critical that the ultimate user of the system "buys into" the model methodology. By using an interactive approach the planner still plays a key role in generating a schedule. The user does not relinquish total control of his job function to the system, but is supported in his decision-making efforts by allowing the machine to handle the computational burden of generating a plan. This elevates the scheduler from the level of a number-crunching "technician" to a "maintenance planner and analyst," freeing him to think about better ways to solve the problem at hand and to do exception handling, which makes better use of the human's rational capabilities.

This methodology provides a vehicle for the planner to use his experience and knowledge of maintenance planning to guide the model's scheduling heuristic in the right direction toward creating an optimal dock plan. The user will be much more likely to use a system that he fits into and understands. Later, the user will also be likely to accept a more complete and total system solution, if and when one is developed, as a next step in the model development process.

Overhaul Maintenance Model

The aircraft overhaul maintenance scheduling problem can be defined as a job-scheduling-on-parallel-machines problem with precedence, deadline and machine utilization and availability constraints. Fortunately, the overhaul maintenance scheduling problem has an inherent structure which can be exploited to devise a relatively quick and near-optimal heuristic solution procedure. This structure will simplify the job-scheduling-on-parallel-machines problem to a more easily solvable scheduling problem by significantly limiting the possible number of job ordering and assignment permutations. The resulting problem ends up being very similar in structure to an assignment problem and is solved accordingly.

A natural overhaul sequencing mechanism arises from the fact that each overhaul check must be performed prior to reaching its allowable limit. An overhaul' s "allowable date" corresponds to the day in the planning horizon that the aircraft's check will reach its allowable limit and, hence, must be performed. It follows that aircraft overhauls which have accumulated more hours (i.e., are *closer* to *their allowable limit*) should be processed before aircraft overhauls with fewer hours. Sorting overhauls in descending order on hours accumulated against the allowable limit thus provides a logical ordering of overhaul checks (this is equivalent to sorting checks in order of allowable date, with the earliest date first). This sequencing mechanism also contributes towards meeting the objective of yield maximization, as it would be counterproductive to service an aircraft that has less yield (i.e., fewer hours) than one which has greater yield.

Given a sequenced list of overhauls to be performed, an assignment mechanism must be established to assign overhauls to dock lines in some fashion. In order for an overhaul to be assigned to an overhaul dock, the dock must meet certain minimal or candidate criteria, such as being able and available to perform this type of overhaul on this type of aircraft during the time frame that the overhaul may be scheduled.

The "optimal" dock line to which an individual overhaul activity should be assigned is the one that best meets the following criteria:

- aircraft overhaul begins prior to its allowable date;
- aircraft overhaul's start date is scheduled such that its yield is at a maximum between the upper and lower yield limits;
- the overhaul is compatible with the previous program scheduled in this dock; and
- if there exists a preferred dock in which this overhaul should be performed, and all other criteria are met, then the preferred dock is selected.

Once an overhaul has been scheduled, the model generates or "spawns" a successor overhaul for this aircraft. An allowable date for this aircraft's subsequent overhaul is calculated according to the associated allowable limit parameter and the aircraft's utilization rate. This check will be scheduled at a later date, assuming that it falls within the planning horizon. The model terminates when all overhauls in the 5-year planning horizon have been processed.

Heuristic Approach

The need for a heuristic approach versus a strict formulation solution or optimal algorithm approach is justified vis a vis many circumstances specific to this scheduling problem. The exact amount of dock line capacity to have open at each point during the planning horizon is rarely, if ever, known prior to running the scheduling model. The delicate balance of optimizing yields so that no aircraft exceeds its allowable limit depends heavily on having the right amount of capacity available throughout the planning horizon. If an overhaul could not be scheduled due to insufficient capacity, an optimizer (e.g., MPSX or MPS-III) would return an infeasible solution. In the strictest sense this is true; however, the user can modify the amount of capacity available and rerun the model to easily correct the infeasibility.

On the other hand, excess dock capacity leaves gaps in the overhaul schedule, and would also result in an infeasible solution due to the inability to satisfy the constant dock utilization constraint. Similarly, if an overhaul could not be scheduled such that its yield falls between the lower and upper yield control limits, an optimizer would, again, return an infeasible solution. Such a solution, however, may be acceptable to the user in some instances.

Due to the unique structure of the overhaul scheduling problem in which checks may be sequenced on allowable date deadline, the greedy heuristic performs quite well. The algorithm uses the deadline to provide a natural ordering of checks, selects the check with nearest deadline, and positions that check in a dock line such that its yield is as large as possible. The quality of the solution is highly dependent on the user's skill at specifying available dock line capacity and using the yield control limits to drive the solution in the right direction. Upon implementing the algorithm, it was discovered that given a good starting point, the algorithm does quite well at achieving yields between 90 and 100 percent of the allowable limit on most checks.

The use of a greedy, quick heuristic is justified based on the interactive optimization approach to achieving "good, feasible scheduling solutions."

The use of a greedy and relatively quick heuristic is also justified based on interactive optimization approach to achieving "good, feasible scheduling solutions." A considerable amount of interaction between the user and model is necessary to generate an acceptable schedule. The iterative development process is made faster and more productive with the use of a straightforward, easily understood heuristic algorithm, instead of a more rigorous optimization approach.

Benefits of DockPlan System

American Airlines has derived three primary benefits from DockPlan since production usage began in October 1991: 1. planner productivity improvement, 2. maintenance cost avoidance (and reduction), and 3. revenue generation opportunity identification. These three benefits have come from a re-engineering and automation of the existing dock plan development process and an enhancement of the dock plan product itself.

The best evidence of the benefits of DockPlan is its run-time performance. The following performance statistics provide examples of 5-year dock plan development turnaround time:

- A DC-10 fleet (40 aircraft) 5-year dock plan for main base visits and component removals can be generated in about one minute.
- A 727 fleet (164 aircraft) 5-year dock plan for heavy and light C checks and component removals can be generated in about five minutes.
- A Super 80 fleet (250 aircraft) 5-year dock plan for heavy and light C checks and component removals can be generated in about eight minutes.

Automation provides dock planners the capability to generate and evaluate several sub-fleet level "what if... analyses" in a single day. The previous manual process required several hours, and in some cases, days, to generate and fully evaluate a single dock plan scenario.

The capability to optimize yields by better controlling maintenance facility capacity allows maintenance planners to use DockPlan to significantly reduce total overhaul maintenance costs. Recent experience has shown an average increase of 15 percentage points in yield, from 80 to 95 percent of allowable hours, on widebody aircraft heavy C checks. This increase translates into avoiding, on average, two heavy C checks in the life of each widebody aircraft in the fleet. At $1 million per heavy C check, this equates to a potential $454 million overhaul maintenance cost avoidance over the active life of the 227 widebody aircraft in the fleet.

In a recent cost avoidance scenario, dock planners identified six months of excess 767 overhaul maintenance capacity at AA's new Alliance-Fort Worth Maintenance Base, using DockPlan's yield and resource utilization optimization capabilities. This excess capacity was used to perform 767 airframe conversions that were previously planned to be contracted out to an outside vendor maintenance organization. Bringing this work in-house saved AA over $3 million in maintenance labor costs.

In a recent revenue enhancement scenario, dock planners identified an opportunity to shut down a 727 overhaul line one year earlier than expected and still complete all the necessary checks, within their allowable limits, for that fleet. This equates to giving an aircraft back to the airline for an entire year which would have normally been out of service, and hence not available to generate revenue. This result will provide the airline with a considerable increase in revenue generation potential in 1992 than it would have had without DockPlan.

Douglas A. Gray is a principal with American Airlines Decision Technologies in Fort Worth, Texas.

"Airworthy," OR/MS Today, December 1992.

6

Consulting Concepts Learned from Airworthy

Introduction

I wrote this article upon reflection on the project described in Chapter 5.

Fortunately, a lot of things went right on the AA heavy maintenance check and hangar planning and scheduling project, and the end result was a success, as evidenced by many favorable economic impacts and business value generation outcomes. I wanted to understand and capture what went right in an effort to define a *repeatable consulting process* for O.R. projects. These lessons learned and the process still hold true today, more generally for analytics, data science, and AI projects.

The key attributes in the consulting process include:

- Interdisciplinary nature of the project with a variety of business stakeholders, subject matter domain expert constituents, O.R. analyst, industrial engineer, and software engineer
- Role of the O.R. analyst/consultant
  - Industry knowledge acquisition
  - Understanding the underlying business process, data, objectives, variables, constraints
  - Understanding the underlying business problem and organizational objectives
  - Providing decision support for decision-makers
- Holistic approach to problem solving
  - Interpersonal and communication skills, e.g., "people skills," emotional intelligence (EQ)
  - Technology and modeling skills, i.e., recognizing a problem and applying the proper model and algorithmic solution approach, then coding and presenting the results in a manner that the end users can understand and apply; IQ

 DOI: 10.1201/9781003588344-7

- Understanding the client's real needs to solve the problem
- Gaining client acceptance and support (aka, "winning the hearts and minds")
- Importance of champions and motivated sponsors who have a "target" KPI in mind, i.e., check yield and maintenance check costs, that is realistically quantified and achievable
- Importance of a project "plan" aka proposal, work plan, charter
- Client participation and engagement to ease the pain of change management

How does this consulting process compare to your approach to analytical sciences projects?

BROADEN PERSPECTIVE: CONSULTING CONCEPTS COME TO LIFE FOR AUTHOR WHILE WORKING ON AMERICAN AIRLINES MAINTENANCE PROJECT

The application of operations research in business, government, public or private sectors requires a multi-faceted approach to be successful. This article introduces a consulting process which is applicable in any environment for the development and implementation of OR model-based decision support systems (DSS), where a DSS is defined as any computer-based system which assists users in some decision-making process.

The management consulting technique of identifying a client's needs and meeting those needs with solutions that provide benefits is one which is fundamental to OR analysts. This technique, coupled with the recognition of the OR consultant's role in decision support, is critical to the success of this OR consulting process. Included in the process are the necessary system components which mirror typical decision processes, such as data development, what-if analysis and quantitative decision evaluation. Also included are the supporting elements which must be in place in order to successfully implement the system with the client-user group, such as client participation, system training, support, documentation, as well as a fundamental understanding and relationship between client and consultant.

This technical consulting methodology is founded on the problem-solving and interpersonal skills of the OR consultant, both of which are necessary to build long-lasting client-consultant relationships. However, the use of modern technology, such as graphical user interfaces and rapid prototyping, in addition to traditional modeling approaches, provide new dimensions in which OR consultants can positively impact the organization and the bottom line.

Although successful in several applications at American Airlines, the methodology presented here is not intended to be a panacea for the application of OR, but more an example of how OR can be successfully applied through a consulting process.

REAL-WORLD PROJECT

In order to solidify and give credence to methodology, specific examples are included from an actual OR model-based system development project recently undertaken by the author. This project, which involved the development of an aircraft overhaul scheduling system for the American Airlines Maintenance and Engineering Division, is a particularly good example on which to focus for several reasons. It represents a complete systems development process from system conceptual design through development, implementation and support, and demonstrates the enormous benefits that can be achieved when OR is applied with the client's best interests in mind. In fact, it was the success of this project that motivated the author to write this paper, and thereby formally recognize the consulting concepts learned during this process.

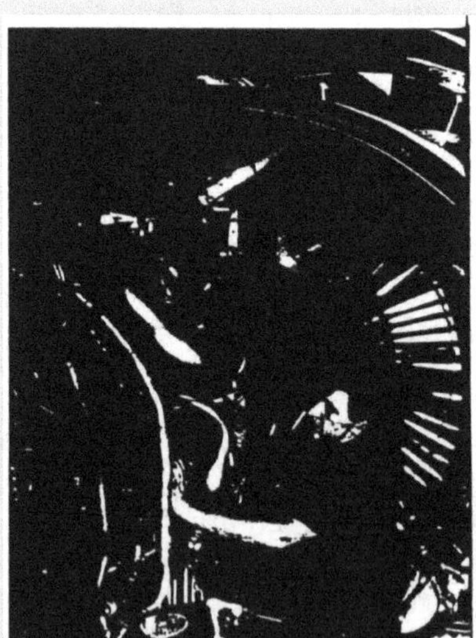

To be successful, the process by which OR is applied should be similar to the nature of OR itself, i.e., interdisciplinary. It should also reflect the nature of the clients it attempts to serve, i.e., results-oriented and people-oriented. By employing a mixture of concepts from OR, management, computer science, software engineering, business, psychology and marketing, a complete consulting process for providing effective decision support and solving problems in business, industry, government or the public sector can be developed. An approach which emphasizes the importance of interpersonal and "people" skills as much, if not more, than technical, problem-solving skills is a necessity.

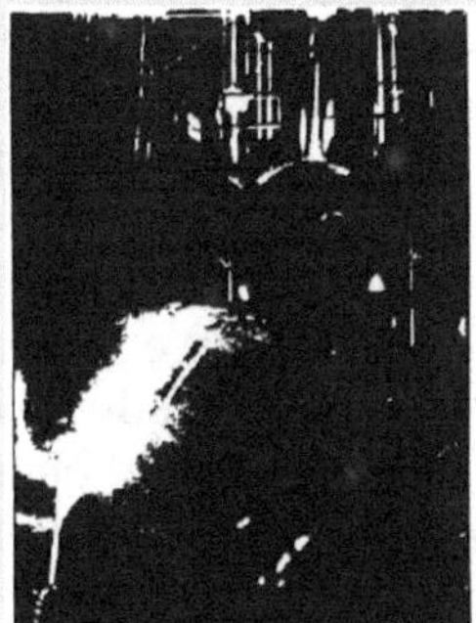

The project: develop an efficient aircraft overhaul scheduling system.

The clients we worked with to develop the aircraft overhaul scheduling system are a part of the Long Range Planning Group of the Maintenance and Engineering (M&E) Division based at the American Airlines Maintenance and Engineering Center in Tulsa, Okla. An interdisciplinary approach to the systems development project was taken from the start involving a maintenance scheduler (the ultimate user of the system), an industrial engineer supporting the maintenance schedulers (who acted as a liaison between the consultants and the client user group), an OR consultant (i.e., the author, responsible for problem analysis and scheduling model development), and finally a systems consultant (responsible for developing the input/output user-interfaces and report writers).

The systems development activity was motivated by a cost justification, performed by the industrial engineer in the client-user group, which identified a significant potential annual cost savings which could be achieved if such a system were developed and implemented. However, due to the complicated nature of the problem, the corporate environment and an off-site development, close contact and coordination between clients and consultants was necessary to successfully develop and implement the system.

THE ROLE OF THE OR CONSULTANT

OR consultants, to be effective, must understand their role within and relative to the organization they are serving. First, they must develop a comprehensive knowledge of the industry or business that the organization is in, be it airlines, railroads, steel-making, health care or public service. Secondly, OR consultants must understand the objectives of the organization so that they may employ their skills and abilities to better serve them. Finally, they must embrace their role of decision support, that is provide decision makers with analyses and systems that provide an objective look at alternatives, potential benefits and consequences.

The first step in the system development activity was for myself, the OR consultant, to learn and understand everything there was to know about aircraft overhaul maintenance scheduling. I needed to know not only the way in which the clients did business today, but also how they envisioned doing business with the new system. This involved spending several weeks working with the clients on-site and remaining in close contact to answer any and all questions that arose. The objectives and constraints of the problem as well as those involved in the systems development activity itself were researched, defined and accepted by everyone involved in the process. As consultants, we embraced our role of decision support by trying to suggest and develop ways of making better bottom-line-oriented decisions with the new system and its scheduling methodology.

UNDERSTANDING CLIENT'S NEEDS

Too often, OR practitioners (and academics) look strictly at the technical aspects of a problem – a model formulation, or the technique used to solve a particular model – and miss the point as to why the model was needed in the first place.

Therefore, understanding a client decision-maker's needs and the nature of their business which underlies the decision process or problem at hand is critical to providing a viable solution. Understanding a client's business is of great importance in gaining the client's acceptance and approval of the model or system which the OR consultant will produce.

Gene Woolsey has said for years that "if a consultant doesn't understand the way their client does business today as well, if not better, than they do, and then attempts to tell them to do business another way, then the consultant is a fraud."

The client's primary objective for the new system was to produce overhaul schedules which better utilized available hangar capacity and reduced the total number of overhaul checks performed, while remaining within government regulatory aircraft maintenance guidelines. However, the client's needs for the system's functionality included the ability to:

- *Quickly generate aircraft overhaul maintenance schedules from input data.*
- *Perform what-if scenario analysis on the large number of changing variable and parameter values involved in the maintenance scheduling process.*
- *Analyze and evaluate the quality of a schedule in comparison to other schedules.*
- *Efficiently maintain and manage the data associated with aircraft overhaul maintenance scheduling.*

The understanding of these goals and needs allowed the system specification and development process to proceed with few unforeseen circumstances. In fact, using available technology, we, the consultants, were able to far exceed the minimum requirements put forth by the user and provide them with imaginative as well as functional solutions and capabilities.

Understanding a client's needs and the nature of their business is critical to providing a viable solution.

ATTITUDE

Often, clients harbor feelings of intimidation or ill-will with regard to OR consultants. Several factors cause this type of behavior:

- OR consultants are usually better educated than most of their clients in business.

- OR consultants are viewed as "cost cutters" and "job killers."
- Clients sense OR consultants will "expose the deficiencies" in the status quo organization.
- Sometimes OR consultants incorrectly portray themselves as having superior knowledge and experience in the situation at hand.

To be successful, OR consultants must actively work to gain the acceptance and support of their clients. They must convince the clients that they are there to help the clients perform their tasks; not to expose their failings or recommend their termination. They must convince their clients through their actions, not words, that they want to "work with them to improve the situation"; not to "dictate the way the world really is."

From the very beginning, we attempted to win the acceptance of the clients by listening to what they had to say, even if it was not directly relevant to the specific task at hand. We assured them that no computer solution would replace the insight and expertise that the human scheduler brings to the scheduling process; rather, that we were attempting to build a system which would support their decision making by helping them do their job more efficiently and more successfully than ever before. We communicated that it was our goal to build a system that would save American Airlines several millions of dollars in aircraft maintenance costs every year by significantly improving the overhaul maintenance scheduling process.

CHAMPIONS NEEDED

Tom Peters, the well-known management consultant, has said, "Anytime anything gets done anywhere in business, it is because of a champion." A champion is defined as the person who leads the effort to initiate a project, fights the battles, defends the project, refuses to let the project die, and pilots the project through to a successful completion.

The acceptance and implementation of a "champion-of-the-cause" concept in the world of OR consulting is long overdue. Too many times projects are begun with good intentions on both sides, but because of lack of support, or lack of focus and direction, or political consequences, the project is intentionally scrapped or inadvertently left foundering.

Champions are needed on both the client and consultant sides of the project. A project's potential for success is considerably higher if there is a champion on the client side; someone who "knows the players, politics and power centers" and who can successfully support and maneuver the project through the bureaucratic and political minefield that has doomed so many OR projects. Even stronger support can be gained if that client champion has some understanding of the consultant's methods, organization and potential benefits that the OR consultant can provide (like an industrial engineer or quantitative MBA). The client champion must also be willing to make "a leap of faith" and support the notion that the resources spent by the OR consultant will achieve the desired outcome and realize the expected benefits.

The champion on the OR consulting side must work to ensure that the distance that the client champion must leap is minimal. Gaining client faith and support happens through a mutual understanding of the project's goals, objectives and expectations, and frequent communication on the status of the project. The OR champion must sincerely communicate interest in, and enthusiasm for, the project and make sure that the client truly "buys in" to the techniques and methods to be used. The OR consultant must speak the client's language and explain methodology and expected benefits in terms that the client will understand.

A trust must be established between the client and consultant to ensure the former's expectations are realistic and that the latter is delivering as promised. Honesty on the true status of the project is of paramount importance. If the project is off schedule because of delays or unforeseen obstacles, it is better to be candid and inform them accordingly. If the delays are of a reasonable nature, the client should be willing to accept them.

The industrial engineer, working as a liaison between the client and consultant groups, was most definitely the champion on the user's side. Whenever interest waned or doubts were raised about the viability of the system or its potential benefits, this person charged to the rescue and fought to keep morale and interest high. Through an understanding of the client's needs and objectives and the consultant's potential contribution, the user champion was able to keep the momentum going despite difficulties along the way. The consultant champion (the author), through a thorough understanding of the client's business, built a system which directly addressed the client's needs and generated benefits far beyond those initially expected, helping to drive the project to a successful completion.

To be effective, OR consultants should develop a comprehensive knowledge of the industry or business they are involved with.

One monument to the communication between client and consultant that made the project successful: The client organization underwent two middle-level management changes (almost certain death for any corporate endeavor) during the course of the project, yet the project stayed afloat. Senior-level management remained aligned throughout the middle-level shake-up which helped keep the system development process on an even keel.

DEVELOPMENT PLAN OR PROPOSAL

Every OR project that is undertaken should begin with a written statement of the project objectives, the approach, deliverables and expected completion dates. Such a document, usually known as a proposal, specification or work plan, serves a variety of purposes. The proposal provides a written record of the who, what, when, where, why and how of the project as well as the scope of each participant's responsibilities. The proposal can also serve as a guide or road map for the analysis or system development.

The proposal should be specific enough to hold each party accountable when delivery time comes, but flexible enough to adjust and adapt the project's objectives, direction and time frame to account for unforeseen obstacles and challenges along the way. This flexibility is critical in ventures where there is a significant amount of uncertainty in a project's viability or benefits (e.g., research and development activities).

While no system specification is ever complete or sufficient prior to undertaking the project, such a document is absolutely necessary to avoid misunderstandings of responsibilities or expectations. The system specification should be written and accepted by both consultant and client, prior to beginning a project. Periodic reviews (usually monthly or bimonthly) are recommended to ensure that the project (or system development) is on track and discrepancies are resolved as they are encountered. This tracking process does much to ensure continuous communication between client and consultant and contributes greatly to the project's success.

After the initial research on client needs and desired system functionality was completed, a system development plan was written to provide an estimate of the time and resources that would be required to develop such a system. After several iterations with the client user group, the proposal was accepted. Throughout the entire project, this document served as the primary reference of what was to be delivered and when, and with what level of functionality). Although conflicts and misunderstandings of what was expected and what was actually delivered arose more than once during the course of the project, these differences were always worked out with a solution that was amenable to all parties involved. Neither party ever walked away from the bargaining table, which would have made the problem-resolution process impossible.

Bi-weekly status reports and almost daily phone calls from consultant to client helped keep the project on track. It became evident that sufficient time was not provided in the initial proposal to allow for unforeseen obstacles. However, the delays were seen as legitimate by the client organization and not as excuses made by the consultants. In the final analysis, the project, originally planned as 12 man-months, required 14 man-months to complete, including 1.5 man-months worth of client requested add-on developments.

CLIENT PARTICIPATION

Client participation in and awareness of all aspects of the project is one of the most significant elements in the successful completion of projects at American Airlines. Too often OR consultants discount the opinions, insights and knowledge that a client can provide on how to develop a system or perform an analysis. We mistakenly believe that our education and problem-solving skills put us in a position to overrule what are actually the "realities" of applying OR in business. A better approach is to couple our skills with the client's knowledge and understanding to develop a more complete and informed solution.

Granted, there is a very fine line between having enough objectivity to solve the problem at hand without being jaded by the client's opinions, and yet not being ignorant or indifferent to what the client has to say. Therefore, OR consultants must be attentive listeners, ever-sifting through the content of conversations with their clients, to separate facts from opinions.

Client involvement plays another significant role in the success of a project. Clients who have participated in an analysis or system development are much more likely to accept the results because they take on the pride of ownership. This acceptance and ownership augments the relationship between clients and consultants and goes a long way to ensure the success of the project. A client who feels like part of a team is considerably less willing to quit or abandon the team's end product.

The client organization was very much involved in the project from the very beginning. They recognized the need for the system, documented the reasons why the system was needed, performed the necessary cost justification, collected and developed all of the necessary data to drive the system, and wrote their own system functionality specification. All of this enthusiasm and active participation made the consultants job significantly easier. In fact it is the dream of every consultant to have a client who is motivated to help improve their own situation instead of sitting back and waiting to be presented with "the right answer." Although the client users were skeptical at first of the advanced technology, upon explanation and assurances from the consultants and upon undergoing their own very thorough model verification and validation processes they quickly became convinced that the new solution was, indeed, a better one. It was and continues to be a joint team effort of clients and consultants working together that has made the new aircraft overhaul scheduling system successful.

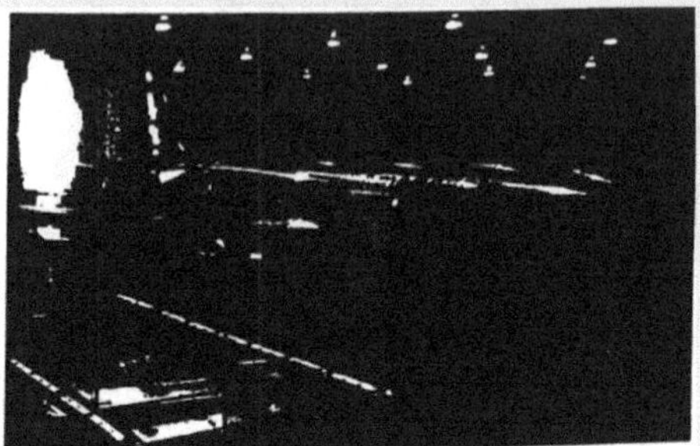

Identified extra hangar capacity will save American Airlines more than $3 million in labor costs.

COMPLETE CONSULTING

Experience has shown that successful OR applications take a more holistic approach to identifying and solving client problems. Understanding a client's business and objectives is a place to start. By understanding their decision-support needs and delivering models and systems that address those needs and provide benefits is where OR can make its greatest contribution.

Instead of focusing on one particular instance of one particular problem in one department, OR must broaden its problem-solving perspective to increase its impact on the organization. In order to "go beyond models" and be effective within organizations, we must develop our interpersonal skills to better communicate and market what we have to offer.

The challenge of the OR discipline in the coming decades will be to employ a complete OR consulting process that uses technology effectively in conjunction with our people skills to positively impact business and industry.

Recognizing M&E's Long Range Planning's function as a business – and treating their maintenance schedule as the product that that organization produces – led us to our purpose of developing a system that would allow planners to create the best maintenance schedule product that was possible in an efficient and timely fashion. By being sensitive to the client's needs and implementing appropriate levels of OR and computer technology, a system solution was developed and successfully implemented that will continue to serve the needs of maintenance planners for several years to come.

The degree of improvement in the aircraft overhaul schedule development process was recognized when a planner identified six months worth of excess *overhaul hangar capacity using the new system. This extra capacity, if utilized to perform aircraft conversion work that would have otherwise been contracted to an outside vendor, will save American Airlines over $3million in labor costs. Improving the methodology and process by which decisions are made and providing users with the right tools to identify such cost saving opportunities is most definitely where OR can make its greatest contribution. We would like to believe that we help people to, in the words of American Airlines President and CEO Bob Crandall, "work smarter, not harder."*

Douglas A. Gray is a principal with American Airlines Decision Technologies in Fort Worth, Texas

"Broaden Perspective: Consulting concepts come to life for author while working on American Airlines maintenance project," OR/MS Today, December 1993.

7

A Modern Day Project Applying the Same Principles: Advanced Analytics Commodity Case

Introduction

This article has two important connections to other work previously mentioned. First, I posted this article as a follow-up and concrete example of how advanced analytics can be *economically transformational*, in practice, in the real world, as described in Chapter 4. Second, it is a more recent implementation of the same type of technical approach and consulting process outlined in Chapters 5 and 6.

There are a handful of key takeaways:

- How the larger problem was decomposed into four components, each solved separately, but the components interlock
- The sophistication of the mathematical modeling-based solution approaches
- How fast the solutions are generated
- The relatively modest level of investment required
- The magnitude of business value and economic impact of the solution
- The strategic, tactical, and operational importance of the solution to the company

Again, we see how important and impactful analytics can be in (somewhat mundane but material) areas of the company, such as purchasing and inventory management.

[LinkedIn]

Following up on a previously posted article on how advanced (predictive/prescriptive) analytics, are economically transformational, and a critical component of digital transformation, I am providing a *real-world example* from my

DOI: 10.1201/9781003588344-8

experience to practically illustrate the thesis. For confidentiality reasons, I disguised the name of the company and their industry, but let's say they are a *Fortune 150* manufacturing firm with revenues of $20+ billion; let's call them *Acme, Inc.*

To produce their finished products, Acme relies on a particular raw material input commodity supplied by third-party vendors; let's call it *gloop*. Gloop is a significant input, and Acme spends billions of dollars per year (20%–30% of annual revenues) to purchase all of the gloop they need to produce their finished goods.

Acme purchases gloop from a variety of national, regional, and local suppliers, and stores the gloop in tanks at their 50 production plant locations around the U.S. The price of gloop fluctuates (sometimes significantly) globally, nationally, regionally, and locally, depending on market supply and demand conditions, and other factors related to gloop production.

There are four parts to the problem of purchasing gloop and managing inventory for same.

First, since the price of gloop varies, which can substantially impact Acme's cost of goods sold, Acme employs commodity hedging using "options" contracts to lock in the future prices of a gloop that Acme will pay on ~20% of the gloop Acme will need to buy (note: options contract costs are prohibitively expensive to hedge 100% of gloop purchases). Like most manufacturers hedging a commodity portfolio, Acme employs a variant on the Black-Scholes options pricing model, along with market data for gloop. Black-Scholes is a multivariate normal distribution that uses partial derivatives from calculus to solve for several variables to determine commodity option values. Black-Scholes can be solved by MathWorks' MATLAB Computational Finance Suite, and is a great example of a *hybrid* predictive/prescriptive analytics model that addresses the *stochastic* nature of gloop prices while *optimizing* against variables, including gloop price, time, option strike price, and volatility. On average, using this hedging approach, Acme achieves a ***net savings of $100 million per year in gloop purchasing costs.***

Second, Acme must forecast demand for gloop on a *monthly* basis at all 50 production plants. Production volume (e.g., make to stock vs. make to order) varies at each plant depending on demand, and is made more variable by seasonality. Demand for gloop varies accordingly. No two plants are alike. Monthly forecasts are rolled up into an annual gloop demand forecast across all 50 locations. (Annual demand forecast information is used to optimize annual gloop purchasing contracts, which are discussed below.) Previously, this process was spreadsheet-based, mostly manual, and wholly unscientific, i.e., rules of thumb. Acme's Analytics Professional and Financial Analyst worked together to develop a machine learning at scale-based forecasting approach to predict demand for gloop in each month (12) at each plant (50) using one or more of 16 different forecasting models (stand-alone vs. ensemble models), including time series (ARIMA, exponential smoothing, moving average), time series regression, and neural networks. These 9,600

(12×50×16) models all solve in ~***five minutes***, and generated gloop forecasts that were ***12%–70% more accurate*** than the prior spreadsheet approach, which took ***three days to complete by hand***. The data were extracted from several enterprise systems using SQL queries, cleansed, and integrated using Alteryx Designer, and all of the forecasting models were developed in R.

Third, Acme must make annual contractual purchasing commitments to their gloop suppliers at each of their 50 production plant locations. Acme leverages the gloop demand forecasts generated in the previous step to inform suppliers of the quantity of gloop required annually at each plant. The contracting process is carried out as an RFP bidding process where suppliers submit bids for quantities to be supplied at set prices at a given plant location. Acme's goal, simply stated, is to satisfy demand for gloop at each plant at minimum total cost. Easier said than done since gloop prices vary by supplier at each plant, and additionally, there is a complex array of taxes, tariffs, fees, and other costs, e.g., EPA charges, that vary at the federal, state, and local levels. The variety of different costs for gloop across 50 plants and 2–3 suppliers at each plant quickly becomes unwieldy, especially because it was previously processed in a spreadsheet. The annual RFP gloop purchasing contract problem is formulated as a fixed-charge mixed-integer linear programming model that simultaneously selects the gloop supplier(s) (in case one supplier cannot supply the amount of gloop required) and satisfies demand such that the total cost across all 50 plants is minimized. Alteryx Designer was again used to extract, cleanse, and integrate the data, and FICO Xpress Optimization was used to solve the mathematical programming problem in a few minutes. Over a three-year period, Acme was able to satisfy demand and ***avoid $38 million in gloop purchasing-related costs*** versus the prior manual spreadsheet-based method. The automated optimization tools also enabled Acme to *experiment* with different purchasing contract arrangements, e.g., *bundling* supplier contracts to fulfill demand at plants in close proximity to one another to achieve volume discounts, which resulted in ***millions more dollars in savings*** on an annual basis. Such experimentation was not possible at all with the prior methods.

Finally, because demand for gloop is *stochastic*, and varies on an *intra-month basis*, and Acme has *fixed* gloop storage tank capacity, it is necessary to regularly monitor gloop consumption and on-hand inventory levels to plan for and schedule gloop replenishment deliveries from suppliers during each month at each plant location. Historically, this process was spreadsheet based, mostly manual, and wholly unscientific, i.e., rules of thumb. The solution approach to the intra-monthly demand and supply gloop inventory problem frames up as a traditional EOQ-Economic Order Quantity model with Re-Order Point, Safety Stock, and Supplier Lead Time considerations. The EOQ model has a nonlinear objective function (quadratic equation), which attempts to minimize a combination of gloop purchasing, shortage, and holding costs, subject to a set of (linear and nonlinear) constraints to account for demand, tank capacity limits, safety stock, and supplier delivery lead times. Demand for gloop during the supplier delivery lead time is modeled as a normal distribution. The EOQ inventory formulation is a *hybrid* predictive and prescriptive analytics model that is solved using a combination of R (normal distribution) and FICO Xpress Optimization (nonlinear mathematical programming model) in *a matter of minutes.* ***Utilizing this scientific approach, Acme is far less likely to unnecessarily over-stock gloop without increasing the risk of a shortage. Having less capital tied up in gloop inventory increases cash flow.***

The multifaceted problem described is solved using a combination of *financial hedging* to offset commodity price fluctuation risk, and *operational hedging* to balance the costs associated with two kinds of *bias*: *financial* bias (minimize cost) and *operational* bias (never run out of gloop!). The inventory problem is *stochastic* and requires *predictive analytics* to forecast demand for gloop, and account for demand during supplier lead times. The problem of minimizing total purchasing and inventory costs is solved with *prescriptive analytics,* in this case, optimization.

Leveraging COTS software, well-known analytical models, and a large volume of data from multiple enterprise systems, one part-time analytics professional and one part-time project manager led by a director, working closely with domain experts, were able to improve Acme's cost structure, cash flow, and profitability. The *benefits* achieved between the *before* and *after* scenarios are significant, substantive, and make a strong case for the economically transformational nature of advanced analytics as a part of digital transformation. Best of all, the 2nd, 3rd, and 4th components of the solution described herein were achieved with a modest investment of ***$400,000 in internal labor costs*** (hardware and software capital expenditures were unallocated and considered part of overhead utilized by many other projects at Acme, Inc.).

8

Right Tool, Right Place, Right Time (with Nader Kabbani)

Introduction

I co-authored this article with Nader Kabbani in 1994 when we were colleagues at American Airlines, based on Nader's experience in flight planning and scheduling solutions, and my experience with airline operations control solutions. Our goal was to characterize planning, scheduling, and operations problems that occur in all sorts of industries by their respective inherent characteristics that mandate a solution approach that is well-suited to handle all facets of each problem's underlying nature. Given our experience, we used airline industry flight planning, scheduling, and operations as the context to explain our approach.

Although the principles outlined in the article still largely hold true today, the advances in computational power of servers have enabled far more robust approaches in solving such problems; for example, "clean sheet" scheduling in airline flight scheduling. Airlines used to start with an existing flight schedule and "tweak it" making small changes to cities served, fleet capacity, etc.; however, clean sheet, as the name implies, enables the airline schedulers to start from "scratch" each time a new schedule is developed, permitting much greater flexibility in cities served, route structures, flight frequencies, aircraft assignments, etc., enabling maximization of airline schedule revenue and profit potential.

One observation worthy of note is the reference to work done in the real-time airline operations control domain at United Airlines that was published in an article entitled "A Decision Support Framework for Airline Flight Cancellation and Delays" authored by Drs. Ahmad Jarrah and Gang Yu, et al. (*Transportation Science,* Vol. 27, No. 3, August 1993, pp. 266–280). The article outlined an approach (i.e., minimum-cost network flow model framework) that was *very* similar to that applied by my team, led by Dr. Mark Song and Dr. Phil Beck at Southwest Airlines, in the development of The Baker, referenced earlier, from 2008 to 2015, and continues there today. (*Small world in airline O.R., as Dr. Gang Yu was Mark's colleague at UT-Austin and at Gang's*

DOI: 10.1201/9781003588344-9

company CALEB Technologies in Austin that won the Edelman Award for their work in Crew Operations Recovery at Continental Airlines; coincidentally, Nader also worked at CALEB briefly after he left AA before going onto a brilliant career at Amazon.)

Alberto Vasquez, John Kirk, Pitu Mirchandani, and myself published a related paper that originated from their work on AA on the Model for Irregular Operations (MIO). [Vasquez, A., Gray, D., Kirk, J., & Mirchandani, P. (1990, May). "A Framework for Implementing Real-time Re-scheduling Systems." *Proceedings, Rensselaer's Second International Conference on Computer Integrated Manufacturing.*]

Airline operating complexity naturally provides an excellent context for understanding how to approach and solve such complicated planning, scheduling, and operations problems.

RIGHT TOOL, PLACE, TIME

Operations planning, scheduling and control (OPSC) problems arise in all sorts of industries, including transportation. These problems are typically centered around the deployment of scarce resources (planes, trains, trucks and machines) among competing activities (flights, routes and products) in such a way that some objective (revenue or profit maximization, cost minimization, or customer service) is optimized, while adhering to operational constraints (regulatory authority-imposed rules, weather patterns and equipment limitations).

Often times, complex interrelationships exist among and between activities and resources that are of a temporal spatial precedence or conditional nature. For example, a particular activity must be carried out within a given time window (customer delivery must occur between 10 a.m. and 2 p.m.); available space in the aircraft's cockpit limits the number of mechanics who can physically fit there to perform maintenance tasks; cartons must be loaded onto a vehicle in the opposite order that they will be delivered; a certain training class can be held in a room that has an overhead projector.

Each stage of the operations management problem – planning, scheduling and control - is recognized by essentially inherent characteristics. These characteristics mandate a solution approach which is well-suited to handle all facets of that type of problem's underlying nature.

Planning assesses demand vis-a-vis resource availability, usually in strategic, aggregate and general terms. This process, sometimes referred to as rough-cut capacity planning, assesses the capability to produce certain products (e.g., petroleum grades) or provide certain services (e.g., flights, line hauls) based on available capacity (e.g., raw materials, planes, trucks). The objective in this stage is to develop a product or service mix which will maximize profit (or revenue) given available capacity and its associated costs, where capacity is a function not only of the number of resources available but also the time frame over which demand is requested for the product mix.

The tools employed in solving planning problems are usually large-scale, optimization-based models and corresponding algorithms. As a result, the problem's inherent characteristics include static, rough-cut data and a lack of extreme time frame limitations (e.g., weeks or months). What-if scenario analysis, sometimes spreadsheet-based, has put a new spin on the concepts of sensitivity analysis in order to provide management planners with the capability to quickly assess the impact of changing parameters on their plan. This stage has the luxury to be forward-looking and proactive in its approach to organizational objectives and is typically not characterized by frenzied, stressful engagements.

Scheduling, more tactical in nature, attempts to create a sequence or order in which activities will be completed, as well as some assignment of activities to available and qualified resources. Albeit a planned assignment, the process considers the characteristics of the activity at hand and the criteria which resources must meet to be considered for assignment to ensure feasibility; i.e., resource qualification and availability. The objective in this stage is to generate a feasible assignment of activities to resources that attempts to realize the profit (or revenue) objectives of the planning stage. Scheduling is characterized by some of the features of planning in that schedules are still relatively static prior to implementation. This stage also embodies some characteristics of operations control that necessitate re-scheduling of activities and re-assignment of resources, vis-a-vis short-term changes in operating conditions such as resource availability or activity duration.

Scheduling is well-known to be a discipline of problems that are computationally intractable in all but a few simple cases and at least hard in many practical cases. Given those realities, the objective is usually to find problem-specific rules-of-thumb or heuristics that quickly provide good, feasible solutions in most practical cases. A combination of optimization-based and heuristic methods are typically employed in scheduling. However, simulation-based evaluative methods, such as what-if analysis, have recently shown promise in scheduling as well [Gray, 1992].

Operations control, as suggested by its name, is a more operational problem and represents where the "plan hits the fan." Operations control attempts to do just that, "control operations" vis-a-vis unforeseen circumstances, e.g., machine or vehicle mechanical failures, economic cataclysms or weather conditions. The objective of this stage is two fold; to minimize the impact of exogenous and sometimes wholly unknown and uncontrollable factors on the planned schedule, and to ensure that the plan is implemented in the most efficient, cost effective way possible to protect the revenue and profit objectives built into the plan.

Whereas scheduling and planning are aggregate, proactive and long-term, operations control is specific, reactive and real-time. This environment mandates less formal and less rigorous solution tools and relies heavily on access to timely and accurate information in a medium conducive to manipulation and ad hoc analysis. Time-honored rules-of-thumb and heuristics grounded in battlefront experience usually help to carry the day. Tools - typically through computer automation - assist operations controllers in monitoring the evolution of the plan, alert them to adverse conditions, and provide information and basic decision support functions to assist in conflict resolution are commonplace.

The graph in Figure 8.1 illustrates the relationship between the level of sophistication of problem-solving tools and the need for timely and

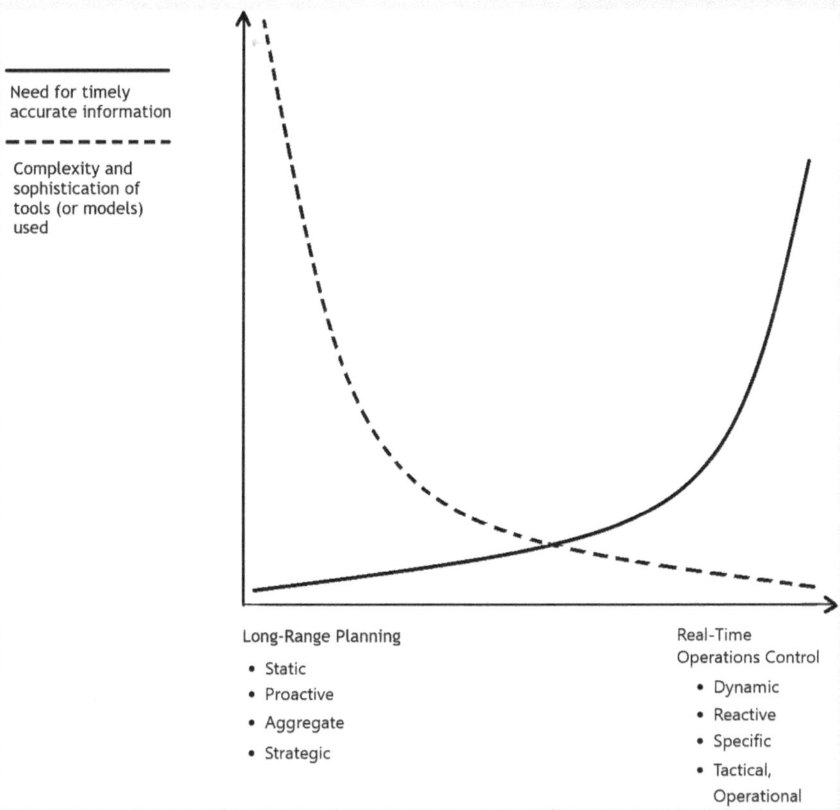

FIGURE 8.1
Operations control, visualized.

accurate information for problem solving (y-axis) over the spectrum of planning, scheduling arid operations control problem environments (x-axis).

While the above definitions of these disciplines should be of no surprise, successful development and implementation of models and systems that accurately address the respective characteristics of OPSC problems is by no means commonplace. Nor are the characteristics of successful application developments widely known or published, let alone regularly practiced. Following are actual project experiences for developing OR model-based decision support systems for OPSC environments.

OVERVIEW

Although OPSC problems are, as defined above, different in nature, their respective solution approaches share similarities. These similarities include a need for:

- data and information (static, long-range for planning and dynamic, real-time for operations control);
- hardware/software computing platforms (parallel computers and optimization packages for planning and engineering workstations and relational databases for operations control);
- graphical user interfaces which provide access to tools (scenario generators in planning and rapid scenario implementation in operations control);
- informal methodologies (what-if analysis in planning and operations control);
- formal methodologies (L/IP optimization for planning, rule- or condition-based monitoring and alerting for operations control).

The challenge is to match the right solution approach to the characteristics of the problem at hand and provide tools that meet the needs of the people attempting to solve the problem. Providing a "black box" L/IP optimizer to an operations controller responsible for running an airline is as great a failure as providing a spreadsheet-based what-if scenario analysis package to a planner trying *to* develop minimum cost flight crew schedules for that same airline. Both are admittedly useful technologies but are hardly interchangeable. Obviously this is an extreme example, but the underlying message is crucial and often missed; that as OR consultants, we must strive to differentiate between our client's planning, scheduling or operations control situation before blindly launching too little or too much technology in the target problem's direction.

AIRLINE FLIGHT SCHEDULING

The challenge of airline flight scheduling provides an excellent vehicle to illustrate the methods and tools employed for decision support in planning, scheduling and operations control. An airline's schedule is the company's sole product and primary source of revenue generation. Major scheduling decisions include the cities to be served, frequency of service and equipment type. The schedule development process is a time-consuming and arduous one, driven by the objective to maximize the profit generation capabilities of the airline's aircraft fleet.

Flight schedule implementation is yet a greater challenge. Once a flight schedule is planned and developed, the objective shifts to implementing the schedule as close to plan as possible, to preserve the "built-in" revenue. Minimizing entropy in, and disruption of, flight schedule operations is easier said than done, considering the myriad exogenous factors that impact airline schedule operation. These factors include weather, mechanical failures and air traffic control (ATC) delays.

FLIGHT SCHEDULE PLANNING

The flight schedule development process begins by developing the initial "skeleton" schedule. Strategic scheduling decisions such as which markets to serve and the level of service at each of these markets are carefully analyzed and made. In evaluating opportunities of serving potential new markets, scheduling analysts practice a very rigorous exercise of analyzing competitive airline schedules, examining new scheduling initiatives, and incorporating new fleet deliveries and aircraft retirements into the new schedule. Similar practices occur in evaluating poor services and deciding on which services to eliminate. During the recent economic conditions, such decisions to streamline services and reduce flying capacity have become very essential survival practices for a number of airlines striving to reduce their unit operating costs.

The main focus during the development of the skeleton schedule is the implementation of market-driven decisions. Such decisions focus on profitability improvements for the airline with considerations of some necessary operational constraints in making these decisions. From the number of complex decisions made during this process, the assignment of aircraft equipment to market segments (legs) has the most impact on schedule profitability. This problem is commonly known as the Fleet Assignment Problem (FAP). FAP is generally formulated and solved as a large-scale mixed integer program.

The problem objectives can vary from maximizing profitability, revenue and fleet utilization to minimizing operating costs. The formulation of such a problem can incorporate a number of operational constraints such as aircraft balancing (flow conservation), aircraft inventory counts, airport, maintenance and many others.

For airlines operating under the hub-and-spoke network, an important scheduling decision to be made is the assignment of through-versus-connecting markets at the hubs. Through markets are ones that continue with the same equipment across hubs, thus creating a marketing advantage to customers compared to connections since passengers do not miss their connections or lose their luggage. This scheduling problem is referred to as the Through Assignment Problem (TAP).

Market-driven decisions made while solving both FAP and TAP are designed to maximize the profitability of the schedule during the planning stages with consideration of the necessary operational constraints. To assure the success of this process, a very thorough exercise of generating and evaluating demand forecasts is used. Market models such as LOGIT or QSI are used to derive market share based on passenger preferences and extensive market surveys.

FLIGHT SCHEDULING

Constraint-based decisions are made during the intermediate-to-short-term flight scheduling. Such decisions are designed to make the schedule operationally feasible while adhering to the impact of schedule changes to the overall schedule profitability. Detailed operational constraints are applied at this stage of the process. Such constraints *vary* significantly and include complex factors such as airport curfews, slots and tower hours; aircraft range, over-water capabilities and seating capacities; maintenance and gating rules and requirements.

Minimizing entropy in,
and disruption of,
light schedule operations is
easier said than done.

One of the most complicated decisions made during this process is the assignment of aircraft routings to meet the airline's operational and maintenance constraints. This problem is referred to as the Aircraft Routing Problem (ARP). The problem is generally solved as a set partitioning problem where a model is used to generate all feasible aircraft routings. Each routing is then evaluated based on its ability to meet the aforementioned parameters. The best set of routings is then selected that meets the airline's operational constraints and has the least impact on revenue. This routing process, being part of the constraint-driven decisions, will attempt to take advantage of optimum fleet assignment (FAP) and through assignment (TAP) decisions made in the long-term schedule planning process (market-driven decisions). If, however, no feasible solution is attained from the existing assignments, the routing process should automatically re-assign either through fleet assignments with minimum impact on revenue to arrive to an operationally feasible solution [Kabbani and Patty, 1992].

Evaluation of the schedule during the flight scheduling process is extremely important to ensure operational feasibility and improved profitability of the new schedule. During this process, the schedule is constantly modified by analysts due to inputs received from a number of sources and departments including fleet planning, maintenance crew, marketing and revenue management. Operational constraints such as aircraft balancing, airport curfews and tower hours, slots, aircraft capacity, range and over-water capabilities are constantly monitored. Similarly, the potential profitability of the new schedule is also evaluated from a revenue, cost and profit prospective. Other airline data is used in the evaluation process to measure the market share of each service offered by the new proposed schedule.

Once the schedule is thoroughly evaluated for meeting all operational feasibility constraints while providing an improvement to overall profitability, the next step is to pass the schedule to the flight operations control for daily monitoring and tracking.

FLIGHT OPERATIONS CONTROL

The problem of controlling an airline's flight operations is inherently reactive by nature. Murphy's Law rules: What can go wrong, will go wrong. Count on it! The approach taken in dealing with this situation is, "What's done is done. Now what are we going to do about it to minimize further down-line disruptions, such as flight delays, cancellations and passenger displacement?"

A flight operations controller's best friend is timely and accurate information about ongoing and upcoming operations. Fast, reliable answers to questions like these are of paramount importance when the "plan hits the fan": "Which aircraft are out-of-service, where, and for how long?" "Where is the storm weather pattern headed, when will it arrive, and how severe will it be when it hits?" "How long are the ATC delays into Chicago-O'Hare and what is the duration of the delay program?"

Obviously, experience – knowing what to do when – is a flight operations controller's greatest ally. But, besides access to and display of information, what types of decision support tools are most useful to flight operations controllers? Four levels of decision support are typically applied in this arena: information processing, operations monitoring and alerting, what-if scenario analysis, and flight re-scheduling and resource reassignment decision support models.

The problem of controlling an airline's flight operations is inherently reactive by nature.

Assuming that a real-time flight data collection, storage, retrieval and communication system is in place, information processing represents a significant part of the flight operations problem solution process. Filtered information displays that allow controllers to quickly assess flight activities at a given airport are invaluable. Such displays provide rapid response to commands like, "List all of the flights arriving at DFW between 0800 and 1000, which are widebody aircraft, with less than 50 passengers." Using these flexible displays, which are based on multiple key searches and sorts, controllers can selectively filter data to support ad hoc analysis and real-time decision-making.

Monitoring and alerting functions allow controllers to use the computer to track flight operations conditions and notify them about situations which are out of kilter. Computers are good at processing large amounts of information quickly, whereas humans are best at exception handling. Rather than having flight operations staff continuously sifting through large amounts of incoming data *to* identify the few situations attention, the computer does the sifting and flags exceptional conditions to the controllers attention.

Using multi-color graphical Gantt chart displays, alert conditions are coded according to problem severity, e.g., all flights which are 30 minutes late are highlighted in yellow, whereas flights more than 30 minutes late are highlighted in red. Controllers can even change the tolerance on disruption notification, since direness of the situation is relative. On some days, 30-minute delays may be typical and three-hour delays are critical. Alerts are also used to flag other noteworthy conditions such as insufficient ground time for baggage and crew connections and violation of airport closure or curfew requirements.

Using a computer model-based representation of the flight schedule network, what-if scenario analysis is employed to simulate flight operations in an attempt to determine the down-line impacts of pending flight decisions. Controllers can specify a set of conditions such as flight delays, cancellations or aircraft reassignments to identify and rectify potential conflicts. Smart simulations supported by databases provide a mechanism to assess the feasibility of flight operations and answer questions such as, "Can this aircraft be assigned to this flight routing which goes over water?" or "Is this aircraft's noise profile suitable to the noise abatement and curfew profiles at the destination airport?"

Experimentation with more sophisticated OR model-based systems that support flight re-scheduling and aircraft re-assignment decision-making has begun to bear fruit. In a recent paper in *Transportation Science,* Jarrah et al., reported on development and implementation of a minimum-cost network flow model framework for supporting flight cancellation and delay decision-making at United Airlines. The model was applied to support real-time operations for United's "hub airports" in Chicago, San Francisco and Denver. The model generated effective, implementable solutions in reasonable time which were in many cases superior to solutions generated by experienced flight controllers in terms of the number and magnitude of flight delays and cancellations required.

Obviously, the level of model sophistication for supporting real-time operations is constrained by the limited amount of time available for solution generation and implementation. However, the rapid advancements and cost-effectiveness in desktop computing power (i.e., engineering workstations), combined with innovative optimization-based heuristic and network modeling approaches, will provide opportunities to apply more rigorous tools for solving such complex, operational problems. The key to the effectiveness of such decision support tools, however, will always be driven by their relevance, applicability and usability in the eyes of the human flight controller, since they have the ultimate responsibility for the decisions that get made and the method employed.

CONCLUSIONS

The decision support tools employed to solve operations planning, scheduling, and control problems are adapted to fit each problem's characteristics and the environment in which decisions are made. While there are significant differences in the level of rigor and sophistication of the OR methods used, there are many aspects which successful applications have in common. These include access to data and information (albeit at varying levels of detail and accuracy), relevance and applicability to the problem at hand, usability of the technology (e.g., through graphical user interfaces), and a capability to experiment using what-if scenario analysis.

The successful OR-based OPSC decision support tools of today - and tomorrow - employ a coherent combination of information technology, software engineering concepts (e.g., object-oriented programming), as well as a healthy dose of creativity and common sense to go along with the sophisticated mathematical models that characterize our profession. The challenge will continue to be to employ the right tools, in the right place, at the right time, with the client and target environment ever in mind to ensure a relatively smooth technology transfer.

Circle #6 on Reader Service Card.

REFERENCES

1. Douglas A. Gray, December 1992, "Airworthiness: Decision Support for Aircraft Maintenance Scheduling," OR/MS Today, pp. 24–29.
2. Nader Kabbani and Bruce Patty, "Aircraft Routings at American Airlines," *Proceedings of the 32nd AGIFORS Annual Symposium, Budapest*, Hungary, October 1992.
3. Ahmad I. Z. Jarrah, Gang Yu, Nirup Krishnamurthy and Ananda Rakshit, August 1993, "A Decision Support Framework for Airline Flight Cancellation and Delays," *Transportation Science*, Vol. 27. No. 3, pp. 266–280.

Douglas Gray is a principal and Nader Kabbani is a senior principal with American Airlines Decision *Technologies in Fort Worth,* Texas.

"Right Tool, Place, Time," OR/MS Today, April 1994.

9

Under Fire: Lessons from the Front – Revenue Management for Manufacturing

Introduction

I wrote this article in 1994 in an attempt to relate the concepts of airline yield management to other industries, specifically manufacturing, in the form of what I refer to as *revenue-based capacity management*. Having seen the revolutionary impact of yield management on airline pricing and seat inventory management – post-U.S. airline industry deregulation in 1979 – I thought there might be an opportunity to apply these concepts, models, and technology to other industries *outside* of transportation. Beyond airlines, yield management has been widely successfully applied to hotels, cruise lines, rental cars, tours, railroads (passenger and freight), and even self-storage companies. All of these industries share the basic characteristics suitable for yield management, i.e., perishable commodity product inventory supply, variable (low brand loyalty) consumer demand, and highly competitive market pricing.

The semiconductor ("chip") industry uses a form of yield management (YM) to decide how many of each type of chip product to make subject to market demand forecasts, product profit margins, and volatile product yield from the manufacturing process. Today, there is plenty of literature, and even commercially available software products, that address YM in semiconductor manufacturing.

At the time, I had the opportunity to do a small advisory project for a U.S. paper products manufacturer that wanted to emulate and apply the types of systems that airlines use for their paper products manufacturing plant production planning. For the article, I utilized a hypothetical paper products company as a contextual template for how to apply *revenue-based capacity management*. It was a nice theory, at best.

I personally never worked in manufacturing, so I never had the opportunity to *apply* these concepts myself. However, 25 years later, when I was teaching Business Analytics at SMU in the EMBA program, a student, who was a deputy superintendent of a steel mill (making ingots, pipe, rebar, and

DOI: 10.1201/9781003588344-10

wire), applied these general concepts in the context of product mix-based production planning utilizing a combination of product-level demand forecasting and linear programming to optimize his manufacturing operations and determine how much of each product to make and when to maximize steel mill profitability. He considered product-level demand and production costs, including setup, manufacturing capacity operating costs, and raw material costs, and product-level contribution profit margins to *increase his plant's profitability by 23%.*

It was incredibly rewarding for me as an educator to see these concepts come to life – *25 years after I had written the article* – and be successfully applied by one of my students working in manufacturing. *His approach had worked.* He received a promotion and was asked by his division GM to apply his model to *every steel mill in the division. Proof point!*

Today, Googling "yield management in manufacturing" results in several scholarly articles on the subject and a definition:

Broadly defined, Yield/Revenue Management is to sell the right inventory or capacity to the right customer, at the right time, and at the right price.

_ Maybe there is something to this concept after all!

UNDER FIRE: LESSONS FROM THE FRONT

The deregulation of the U.S. airline industry in 1978 thrust airlines into competitive warfare. Visionary senior management at companies like American Airlines invested in sophisticated information and decision support technologies to optimally allocate revenue generating resources, and to gain a strategic competitive advantage.

Revenue-based capacity management systems were developed for allocating aircraft to routes and seat inventories to fare classes, vis a vis passenger demand, which optimize revenue generation potential.

In the 21st century, U.S. manufacturing companies will continue to be besieged by significant global competitive pressures. Companies need to understand and anticipate market demand and effectively allocate all resource capacity, i.e., plant and human capital, to ensure survival, let alone profitability. The concept of revenue-based capacity management, which aims at satisfying customer demand by allocating resources so that revenue and profitability are optimized, has been cultivated in the airline industry and shows promise for application in manufacturing- related industries.

U.S. AIRLINE DEREGULATION

Deregulation of the U.S. airline industry in 1978 was clearly a watershed event in the history of U.S. transportation, and in particular for American Airlines (AA). This single event marked the beginning of fierce, unbridled competition among domestic air carriers, the likes of which was unparalleled in any U.S. industry. AA responded by establishing a forward-looking corporate strategy characterized by expansion, product and service innovations, and strategic applications of advanced information and decision support technology.

The objective of providing a wide range of affordable air transportation services to the general public has been duly achieved via deregulation. Under the new rules, airlines are able to fly where, when and as often as they choose, charging fares established by competitive market forces. However, in a capital-intensive industry such as air transportation, where resources and facilities create enormous operating expenses, deregulation created a set of competitive challenges to ensure profitability in what previously had been a lucrative, regulated industry.

As with any set of challenges, there exist a corresponding set of opportunities for those with vision and who are willing to invest, take the chance to compete and win. Although no single event like deregulation has occurred to shock manufacturing companies into action, U.S. firms have been besieged and will continue to be affected by fierce global competition. The following presents the motivation for revenue-based capacity management, as originated in the airline industry as a direct result of deregulation, and addresses how this technology could be transferred to manufacturing companies to create strategic competitive advantage.

LOGISTICAL NIGHTMARE

Operating an airline is a classic example of a logistical nightmare. The already arduous task of scheduling flights and crews is further complicated by unforeseen factors, such as weather and mechanical equipment failures. In an industry where resource and labor costs are high and profit margins are low (1%–2%), airline companies quickly realize the absolute necessity of well-planned and well-executed operations to ensure a profitable enterprise. Based on the performance of commercial airlines recently, well-planned and well-executed operations do not guarantee profitability. However, without logistical cohesion, the commercial airline operation is doomed to certain failure.

Deregulation created a competitive environment that forced AA to manage operations efficiently and cost effectively. Even more important was the emphasis on allocating assets in a way that ensured optimal revenue generation potential. Deregulation motivated AA to harness the power of technology in nearly all facets of airline planning and operations management. Leveraging the vast data resources of AA's SABRE™ Computer Reservations System (CRS) and related systems, the airline was able to deploy information technology coupled with OR/MS model-based decision support systems to augment planner capabilities to make and evaluate decisions.

One outstanding example of airline capacity management is yield or revenue management. The art and science of yield or revenue management is defined as allocating available aircraft seat inventories to fare classes or "buckets" in such a way that revenue generation potential is maximized. AADT won the 1991 Edelman Prize for the best single application of OR/MS for its yield management system, which is responsible for generating an additional 5500 million in revenue each year for AA.

TECHNOLOGY TRANSFER

Yield management is to airlines what revenue-based capacity management is to manufacturers, i.e., how to best allocate available capacity to satisfy customer demand and maximize revenue generation potential. Airlines assign seats to fare class buckets, whereas manufacturers assign production resource capacity, namely plant, materials, and human capital, to product lines, vis a vis customer demand for those products and services.

A yield management system is an airline's single-most important, strategic competitive weapon in the industry's competitive war. It is a system focused on identifying and classifying market demand, creating and pricing products accordingly, and explicitly considering available capacity limitations to maximize revenue generation potential. Similarly, a capacity management system should be the single-most important tool for manufacturers competing globally. Although many companies implicitly consider all of these factors, few have the infrastructure to explicitly link demand and capacity in efforts to optimize profitability.

Although the resources involved are very different, the underlying concepts are the same. Airlines make many of the same decisions that manufacturing companies make every day, e.g., manpower scheduling, job-to-machine assignments. The primary difference between the two industries is the sheer velocity with which fierce competition was brought to bear on airlines, whereas in manufacturing the effects of competition have crept up on and blindsided once profitable, stable companies. This competitive revolution in the airline business forced the airlines to address these challenges and build an extensive technology infrastructure to support activities associated with day-to-day capacity management planning and operational decision-making.

CAPACITY MANAGEMENT

All companies. regardless of industry, must effectively manage capacity to profitably manufacture products and provide services. Capacity is a function of the number of resources available in a given time frame and employed to manufacture a mix of products or provide a mix of services. Capacity is a perishable commodity with which there is associated an opportunity cost for not utilizing a resource to service a customer. Alternatively, there is a cost with overbooking resource capacity, as well as utilizing a resource at a lesser profit margin. Explicitly considering all of these tradeoffs in managing capacity is a formidable task.

Capacity management begins with an in-depth understanding of customer product demand patterns and resource capacity limitations. A fundamental requirement to ensure effective capacity management is a technology infrastructure that supports management of timely and accurate information and decision support on resource capacity implications. Conceptual models that represent customer demand patterns and production resource capacity are valuable in establishing a company's objectives, and the impediments to achieving those goals. The volume of information regarding demand and capacity implications mandates a computer-based infrastructure to manage information flow and support decision-making. Implementation of so-called conceptual models may be in the form of PC-based spreadsheets or engineering workstation-based integrated model-database platforms, or whatever platform is commensurate with the size of the capacity management platform at hand.

American Airlines Decision Technologies (AADT) pioneered the concepts of yield (revenue) management at AA and is the world leader in successfully applying these concepts to other industries, including: hotels, cruise lines, rental car and truck companies, freight and passenger railroads, television and radio stations.

AADT blends state-of-the-art methods and technologies from operations research and computer science to deliver customized capacity management system solutions. Revenue-based capacity management systems could provide a strategic competitive advantage for manufacturers by ensuring that available capacity is allocated with a customer service focus.

REVENUE-BASED CAPACITY MANAGEMENT SYSTEMS

Revenue-based capacity management provides a methodology and an infrastructure for solving the classic OR product mix problem that asks the question: What mix of products optimizes corporate revenue and profit objectives subject to operational and available resource constraints?

The focus of this methodology must be on the customer, i.e., the source of revenue generation. A traditional approach to this problem has been to satisfy customer product demand at minimum cost. An alternative approach seeks to maximise revenue subject to capacity limitations, while controlling costs to achieve profit margin objectives. Key to this approach is understanding and anticipating the marketplace, namely customer needs, wants and perceptions of value, and capitalizing on trends in marketplace purchasing and product pricing. Consistently and accurately forecasting customer demand and factoring it into capacity allocation decisions is critical.

Once demand is targeted, planning production and distribution to deliver products to meet customer demands within limitations of available capacity, follows. This means promising feasible delivery dates, delivering as promised and allocating resources and measuring performance according to meeting customer demands and expectations, e.g., percentage of orders filled within 24 hours versus high machine utilization to minimize production costs.

While effective planning is always a necessity, responding deftly to unforeseen changes in customer demand patterns is even more critical to winning and maintaining customer loyalty. Being able to re-plan and re-schedule production activities, and evaluate revenue and profit impacts of competitive market forces are an important part of revenue-based capacity management. These are important because there is nothing as constant as change in business today. The ability to respond quickly and efficiently can mean the difference between keeping and losing a key customer.

Information empowers the acquisition of knowledge. In business today as always, knowledge is power. The velocity of knowledge is increasing exponentially through the pervasive use of computer and information technology in the workplace. Technology is quickly becoming the single-most important medium of information, and hence knowledge. Knowledge moves quickly, and power shifts accordingly. Competitive power in business today translates into knowledge, i.e., knowing how many to make and sell of what, when, where, at what price and at what cost to ensure profitability. Grasping the power of knowledge depends on two critical factors, namely, availability of information and the capability to harness that information to support decision-making.

Successful revenue-based capacity management systems rely heavily on a technology infrastructure to provide timely and accurate information for decision-making support. An information infrastructure guarantees an information pipeline from the customer to sales and marketing, on-line databases describing products and resource capacity, and real-time data collection regarding factory floor operations.

A decision support infrastructure includes customer product demand forecasting models that employ state-of-the-art methods, such as: exponential smoothing, categorical data analysis and neural networks. Production and distribution planning models employ resource allocation and assignment techniques and finite capacity planning and scheduling to generate feasible manufacturing and distribution plans. These models help to ensure that promised customer delivery schedules are met consistently. Production and distribution operations control models using dynamic re-planning and re-scheduling techniques to support personnel. Computers are programmed to monitor production conditions and alert human controllers to exceptional circumstances that need attention. Real-time decision support systems provide simulation capabilities to conduct what if scenario analysis.

Without an information and decision support infrastructure that is commensurate with the scope of the capacity management problem at hand, attempts at obtaining consistently good, feasible solutions in the face of continuously changing demand patterns and operating conditions is essentially futile.

AIRLINE CAPACITY MANAGEMENT

Although I have focused particularly on yield management, there are three levels of revenue-based capacity management for an airline. An airline must assess customer or market demand to determine where, when and how often to fly. After having established a route structure, the airline must then determine what aircraft types, e.g., wide body versus narrow body, to assign to each route. Finally, once aircraft types have been assigned, the process of yield management begins to solve the problem of deciding how many seats to sell in each fare class bucket so that revenue generation potential is maximized. So, yield management is really a compact form of capacity management, in and of itself.

An airline's product line is represented by a flight schedule, i.e., frequency of flights from origins to destinations, and corresponding flight departure and arrival times. Prices, or fares, charged for these products are represented by discount fares on tickets with restrictions, unrestricted full fare tickets, as well as fares corresponding to different levels of service, e.g., first class, coach class.

Schedules are created based on customer demand for flight products, where a product is defined as a flight from city A to city B, departing at a certain time of day, for a particular fare class. Airlines set fares according to competitive market forces.

The schematic diagram in Figure 9.1 illustrates the "development" or "planning" side of the yield management system at AA. Historical

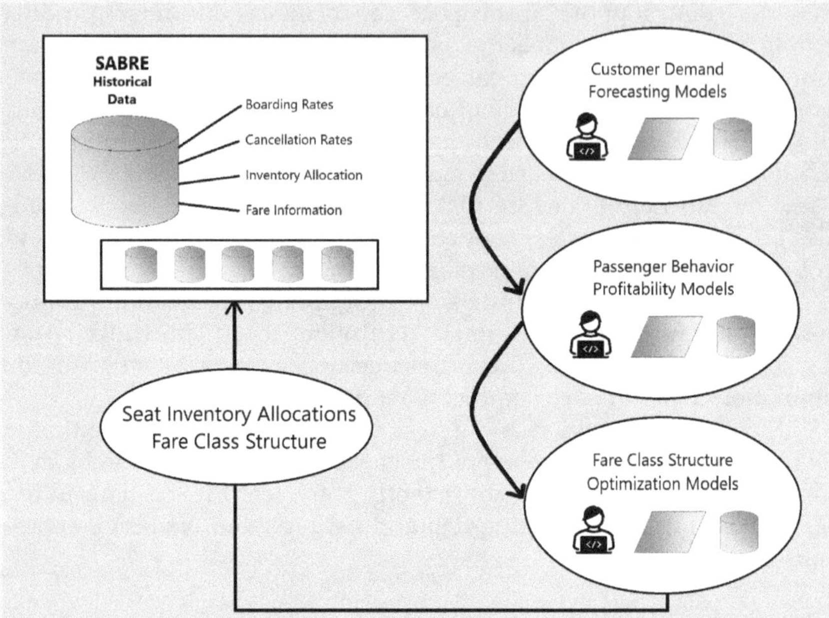

FIGURE 9.1
Airline revenue-based capacity management system – development.

data is retrieved from SABRE™ regarding customer demand for various flight services and corresponding fare information. This data is used to develop and validate a series of forecasting models regarding issues such as flight boarding and cancellation rates. These models assist schedule developers with decisions regarding what type of aircraft equipment should be assigned to flights to best capture customer demand. Using this demand information, stochastic, integer programming optimization models are employed to create an optimal fare structure, based on demand patterns, which will maximize revenue generation potential.

The schematic diagram in Figure 9.2 illustrates the "production" or "operations control" side of the yield management system at AA. As flight day-of-departure approaches, the yield management system monitors customer demand for each fare class, and dynamically allocates seat inventories among fare classes to best capture customer demand and maximize revenue generation potential.

Capacity management in this operation is critical. Selling too many steeply discounted fares eliminates the opportunity to sell a seat to a full-fare passenger who desires to buy a ticket on or near day of departure. Alternatively, if that same full-fare passenger fails to buy a ticket,

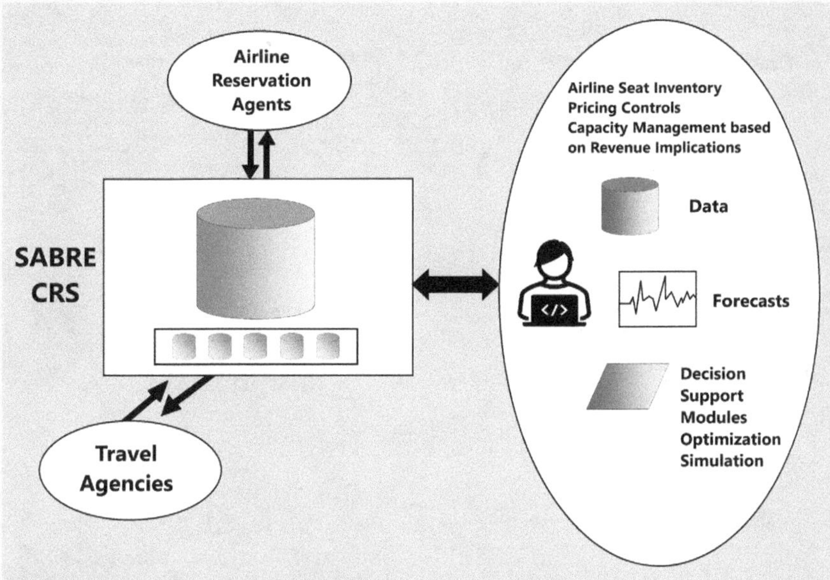

FIGURE 9.2
Airline revenue-based capacity management system – production.

refusing to sell discount fares in advance may eliminate an opportunity to fill that seat, hence losing the opportunity to cover the associated operating cost.

Benefits of revenue-based capacity management for an airline are significant. AA has the lowest involuntary denied boarding rate in the U.S. commercial airline industry. Since airlines overbook actual capacity to compensate for no-shows, there is always a small chance that one or more passengers may not get a seat on the aircraft. For AA, passengers who purchase tickets and show up for the flight are guaranteed, in all but a very few cases, a seat on the airplane.

The airline experiences an annual increase of 3–6 percent in revenue generated as a result of planning and controlling airline seat inventory using revenue-based capacity management techniques. For AA that equates to $500 million annually in additional revenue. Customer service and corporate revenue objectives are achieved simultaneously.

PAPER MANUFACTURING

A hypothetical paper manufacturing company, ABC. Inc., produces paper products for home use, e.g., paper towels, tissues, etc. The diagram in Figure 9.3 illustrates their business process. Marketing

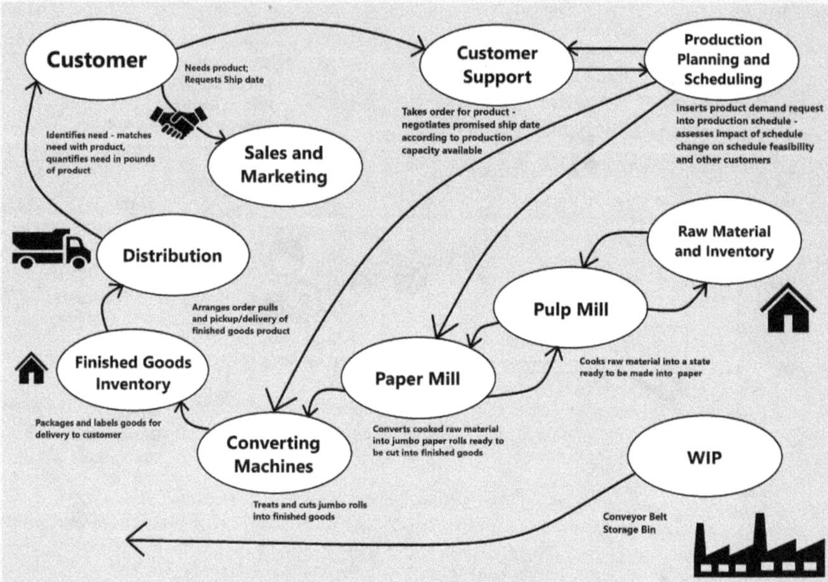

FIGURE 9.3
Paper manufacturing company business process – assess market/customer demand.

establishes high-level customer relationships and customer support administers that relationship by processing product orders. Production planning considers demand for product, coordinates with raw material suppliers, work in process (WIP) and finished goods inventory (FGI), to schedule production and distribution activities to satisfy customer demands. Production coordinates manpower and runs the machines, producing FGI to fill customer orders. Distribution pulls FGI from the warehouse and coordinates pick up with trucking companies.

The schematic diagram in Figure 9.4 illustrates a revenue-based capacity management system vision for the paper manufacturing company. Historical data on individual customer demands are retrieved from the central company database. Trends in demand for product lines are identified using a variety of forecasting models. Marketing staff use these models to better track market forces to anticipate demand to enhance coordination of production activities. Customer support staff is alerted to exceptional conditions, such as a customer who normally orders 10,000 lbs. of product xyz every six weeks who hasn't ordered for seven weeks. Staff can proactively anticipate demand and pursue business rather than reactively responding to customer requests.

The key function in this capacity management system is production planning. Using a combination of material requirement planning

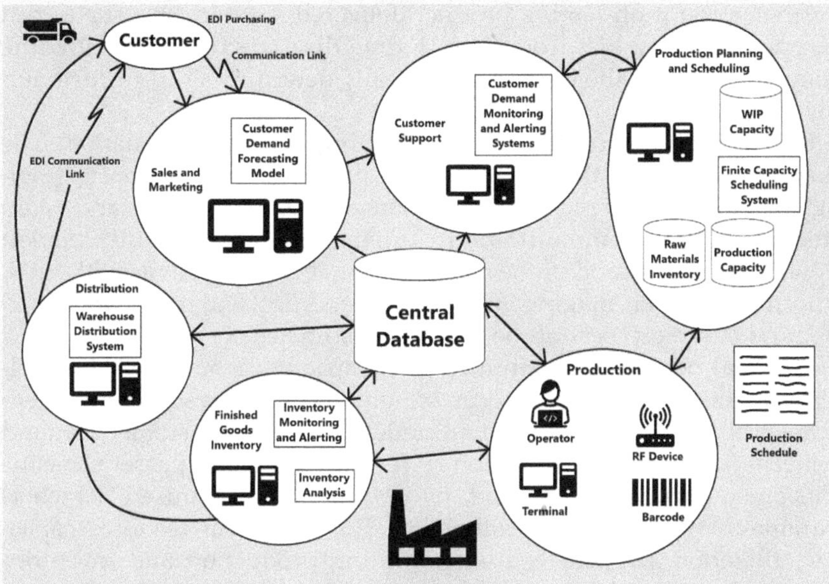

FIGURE 9.4
Manufacturing-based capacity management system vision – assess market/customer demand.

(MRP) and finite capacity planning and scheduling technology, planners can more effectively generate feasible production plans and deftly assess the impact of changes in demand patterns on existing production schedules. This function is critical in promising and meeting customer product delivery dates. A combination of information technology, optimization and simulation modelling augment planners' capabilities to develop and evaluate production and inventory plans: plans that are subject to continuously changing demand patterns, as well as unforeseen conflicts that may exist or disruptions that may occur.

Finally, once production plans are implemented, real-time communication with, and decision support for, production operations is another critical link in the capacity management chain. With little or no slack in the system resource availability and interaction between various stages of the production process, real-time control systems must be in place to monitor operations and corresponding decision support tools must be available to generate and evaluate contingency plan alternatives to ensure that recovery from disruption is quick and efficient.

More importantly, from a customer service perspective, if violation of promised delivery dates is unavoidable, such systems will provide information on why this condition has occurred and when can the customer count on having product delivered. Customers usually can accept the reality that from time to time disasters do occur. How the company handles the fiasco is what really determines future customer loyalty.

Key to these concepts is the underlying flow of information. The diagram illustrates the use of various types of information technology, such as electronic data interchange (EDI), databases, and radio frequency (RF) communication. Timely and accurate information flows are the lifeblood of the capacity management system. Without it, informed decision-making is impossible and decision processes are ad hoc and based on "seat-of-the-pants" guestimates.

Potential benefits of such integrated capacity management technology for the paper manufacturer are obvious: improved revenue generation opportunities through an understanding of customer demand patterns, and optimal allocation of revenue generating asset capacity. Customer service is enhanced, by delivering as promised, which is paramount to sustained profitability. Efficiencies in resource capacity utilization are achieved, streamlining production and inventory requirements, and hence reducing operating and asset costs.

CONCLUSION

Revenue-based capacity management systems have played a significant role in establishing and maintaining American Airline's position as a world-class competitor. No single event, however, has motivated the use of customer- and revenue-oriented capacity management systems in manufacturing industries. Competitive forces, if recognized, will encourage companies to look for ways to leverage available resources to provide even better customer service, while maintaining profitability. Capacity management technology should play a major role in this process.

Revenue-based capacity management is based on understanding customer needs and wants, using customer product demand trends to drive company objectives. Companies must plan production while explicitly considering capacity limitations, using methodologies such as finite capacity planning. Manufacturers must ensure that their production operations are responsive to changes in customer demand patterns and flexible enough to handle disruptions and still meet customer delivery dates. Such responsiveness under reactive circumstances is achievable only if control systems are prepared to handle the entropy common in complex systems, where Murphy's Law is the rule, not the exception.

Revenue-based capacity management systems infrastructure must provide access to timely and accurate information on customer product demand patterns, as well as assessments of production capacity under a variety of scenarios. Decision support tools need to augment forecasting and better anticipating of customer demand, planning operations vis a vis capacity constraints, and monitoring and controlling on-going operations.

While not intended as a panacea to solve all manufacturing industry woes, revenue-based capacity management combines an integrated, customer-oriented approach to resource allocation aimed at optimizing the firm's performance while satisfying customer demand.

In the 21st century world of global competitive warfare, companies that have the vision to invest in developing their technology infrastructure necessary to support revenue-based capacity management will certainly be positioned to achieve a strategic competitive advantage.

Douglas A. Gray is a Senior Principal with SABRE Decision Technologies (SDT), a software development and consulting company based in Fort Worth, Texas.

"Under Fire: Lessons from the Front," *OR/MS Today*, October 1994.

10

Analytics Nontechnical Skills

Introduction

This is a series of three short articles that I posted on LinkedIn between 2020 and 2023, which received a great deal of positive feedback because they address the realities of how the *real world* of applying the analytical sciences is quite different from what is taught in undergraduate and graduate school programs. The second article specifically addresses three common reasons why data science (DS) projects are not implemented, contributing to the 80% of projects that never result in deployment or value capture. Lastly, I address how an MBA has made me a better DS practitioner and leader for those interested in pursuing that type of education.

What They Do Not Teach You in Your MS Data Science/Business Analytics Program

Suffice it to say, DS and analytics are *white hot* career opportunities today, and for the foreseeable future, along with AI. I am encouraged by the upward trend as a classically trained statistician, operations research practitioner, and data and analytics executive with 20+ years of experience solving complex problems within large enterprises, such as airlines and insurance companies, and delivering seven to nine figures in quantifiable annual business value.

Many DS/Business Analytics MS program curricula are largely, if not exclusively, focused on teaching *quantitative techniques*, e.g., machine learning (ML) and predictive models, and *technology tools*, e.g., R and Python. I get it. Faculty are highly quantitatively inclined, and "modeling" is the most intellectually stimulating, easily taught, and "fun" part of doing DS and analytics. Moreover, it is *absolutely critical* for data scientists to understand "modeling mechanics" – such as bias vs. variance, experimental design ("blocking"), and statistical and ML tests for measuring model efficacy (type I and II errors), – to be effective.

That said, after you return your cap and gown and hang your MS diploma on the wall, you are likely to be confronted by *a whole different set of challenges*

DOI: 10.1201/9781003588344-11

executing DS and analytics in the real world – more than your MS degree coursework prepared you to handle. These are challenges that require a different set of "soft skills," which are no less important than your "quant skills."

First of all, the majority of companies are grappling with numerous complex *data* challenges including the following: a myriad of (many times conflicting) source systems, a need for better data governance to enable accessibility, building a data lake, data warehouse, or now a data lakehouse or data mesh to integrate data in one location, and moving to the cloud to handle ever-increasing data complexity and volume. All of these factors complicate and sometimes stymy getting DS projects off the ground.

Even beyond data, there is a whole set of skills and domains that DS and business analytics practitioners need to develop and address to be effective.

- Industry Knowledge
- Business Strategy (Corporate and Departmental)
- Business Processes & Mechanics
- Financial Statements
- Culture and Politics
- Change Management
- Project Management
- Information Technology

Being effective in DS and business analytics requires a natural, inherent curiosity about how businesses work and how they make money. This starts with understanding the industry in which you are working. Oil and gas/energy, retail, food service, airlines, trucking, railroads, financial services, manufacturing, and telecommunications are all very different but use DS to varying degrees of economic impact and success. Within the industry, you need to understand your company's business strategy – what is unique about your firm from a competitive perspective, what makes your company "tick," how does it make and deliver products and services, and how does it make money? Strategy is important at the corporate level, and that filters down into departments. Working in operations is very different from working in marketing or finance – three of the top business functions using DS. Each has its own lexicon and distinct business problems and processes, the mechanics of how things get done, decisions get made, and problems get framed, analyzed, and solved.

Now, I know you purposefully did not get an MBA, but being able to read and understand a company's financial statements, such as balance sheets, income statements, and cash flow statements, yields considerable insight into where a company's revenues are generated and costs manifested – always a great target for analytics – and where the company is making or losing money. In retail, for example, the cost of goods sold (COGS) impacts gross and operating profit margins. In airlines, crew and fuel costs can consume over 75% of annual revenue. Financial statements are a crucially important dialect in the language of business.

A company's *culture* and *political landscape* will heavily factor in determining the ultimate success of analytics, starting with executive, senior, and mid-level leadership. Is the company prone to make *fact-based decisions backed up by data, models, and analysis,* or do the **HiPPOs** rule (**Hi**ghest **P**aid **P**erson's **O**pinion)? If the latter, then *watch out* because when the "data speaks," it usually invalidates outdated assumptions and reveals inconvenient truths about the performance, efficiency, and effectiveness of departments and business processes, which can ultimately threaten the *status quo* of budgets, resources, and power structures. This can be a *political minefield* to navigate, despite your best intentions to improve the firm's economic performance.

Analytics done well begets "creative disruption" by uncovering inefficiency and proposing new ways of doing things better, i.e., more efficiently and/or more effectively, delivering more value and output with fewer or the same number of resources. If and when analytics models are implemented, this type of disruption leads to everyone's favorite topic: *change*. Managing change is critical to analytics success, particularly working closely with business and IT partners from the start. No one likes being blindsided by major disruptive impacts on their part of the firm, no matter how much value is generated with DS. Communication, engagement, and edification are key to managing change, as well as "what's in it" for constituents.

DS and analytics are no different from any other major business endeavor, e.g., developing and implementing a new process or system. The activity, or *project*, must be organized and *managed* – scope, timeline, resources, budget, and quality – engaging constituents on their turf and terms, not in a vacuum or "lab." The principles of project management apply, and are relevant and critical to success, i.e., embedding a model in a process/system for ongoing use. Time-boxing activities with feedback loops is highly recommended to get to a *Minimum Viable Product (or Model)* (in Agile-speak) that is accurate, useful, and generates measurable and substantial incremental business value as soon as possible.

The interplay between DS and IT, and business partners, in big companies is tricky. In many companies, the relationship between analytics and IT is still being ironed out in terms of reporting relationships, responsibility for selecting and implementing platforms, data management, and ownership, and engaging with business partners on projects to implement solutions. There is simultaneously both a distinctive difference *and* overlap between data, DS, and IT scope and purview that necessitates a great deal of communication, cooperation, collaboration, regular interaction, and everyone focusing on what they do best without getting in each other's way.

In the two Business Analytics courses that I teach at SMU (to EMBAs in the Cox School of Business and graduate DS students in the MS program), the above topics are the focus. We leverage Tom Davenport's books (*Competing on Analytics* and *Keeping Up with the Quants*) and frameworks (**DELTTAA,**[1] **FORCE,**[2] **FACE**[3]) applied to real-world situations and case studies to guide leadership and practitioner professionals in developing the "soft skills" that are so critical to DS success at scale in corporate America.

Three Common Reasons DS Projects Don't Get Implemented

Models by nature are experimental and involve scientific discovery, i.e., they do not always work as well as we might like. Beyond a lack of sufficient model efficacy, due to any number of reasons (e.g., insufficient data, excessive complexity, etc.), I have repeatedly observed three reasons DS models do not get implemented:

1. Budget limitations – No different than IT or any other department, DS teams are subject to budget limitations; a particular model may have less business value capture potential than other models, and businesses must draw a line somewhere.
2. Priority changes – DS models must align with business priorities, and those priorities often change for a variety of reasons; an effective model attached to a lower relative business priority can prohibit deployment; also, DS projects (many times implemented as *microservices*) are necessarily dependent on other IT/data applications, each of which has its own backlog and priorities that may not permit implementation of every DS model microservice.
3. Organization changes – Changes in leadership/decision-makers can influence which DS models get deployed; there may be a competing model or system developed by another team that has some edge (performance or political) that supersedes your model; organization changes often drive priority and budgetary changes (see 1 and 2).

While there is no way to entirely avoid any of the above factors, ensuring that DS teams are in constant, close, and clear communication with the business leaders/decision-makers regarding budgets, priorities, and leadership direction can help reduce the number of DS projects that get started but do not get finished or implemented. There are many valid (and many less than valid) business reasons, based on practical realities, why even effective DS models do not get deployed. It is a fact of DS practice. All we can do is endeavor to deliver the maximum business value possible on each DS project while adhering to scope, timing, budget/resource, and quality constraints.

MBAs for Data Scientists (Expensive, but Worth It, in My Experience)

My MBA made me a better data scientist practitioner and leader – counterintuitive? Most people think of DS as a highly technical discipline, which it is;

however, DS *execution* and *implementation* involve many *nontechnical* dimensions. My MBA courses in economics, finance, and accounting helped me to better understand the *economic and financial targets* of DS projects, as well as how to *accurately and correctly measure* impact on financial performance. My MBA courses in strategy and entrepreneurship helped me to better align my DS projects with corporate policy and direction and the impact on competitive advantage. My MBA courses in organizational behavior and leadership helped me to better understand and account for the human dimension, e.g., system and process change management, which DS projects significantly alter and impact. DS practice and leadership inherently require understanding the business from a strategic, process, financial, and human dimension. An MBA can complement the requisite deep math and technology technical skills to help ensure greater success of DS projects.

Notes

1 **DELTTAA** – **D**ata, **E**nterprise view, **L**eadership support, **T**argets (KPIs), **T**echnologies, **A**nalysts, **A**nalytical methodologies

2 **FORCE** – **F**act-based, data- and model-driven decision-making, **O**rganization of analysts, **R**einforcing a culture of analytical decision-making and "fail fast, test and learn," **C**ontinual renewal of business assumptions and models, **E**mbedding analytics in processes

3 **FACE** – **F**rame, **A**nalytically model, **C**ommunicate and act on model results, **E**mbed models in processes and systems

11

Top 10 Analytics Leadership Skills (with Tom Davenport)

Introduction

This chapter presents the research I did based on my own experience and observations in the industry, validated with renowned researcher and author, Tom Davenport, PhD, and posted on LinkedIn. The chapter addresses the top 10 attributes that all analytics leaders should develop and embrace to be effective. As you can see, the majority of the attributes are not at all technical. Managing and leading an analytics group is a lot like running your own business. The endeavor is multifaceted and one in which the technical aspects of the work actually consume a small fraction of the leader's time.

With all of the energy expended and hype generated around analytics, data science, AI, and machine learning, an important question being asked is: *What skills do analytics* ***leaders*** *need to have, or develop, to be successful?*

We describe ten of those leadership skills and traits in this chapter and address what type of professional most likely has these skills. The list may be useful for anyone seeking to hire a leader of analytics or data science functions.

Recruitment, Retention, People Development

The competition for analytics and data science resources is fierce; the qualified resource supply is low; and a leader needs to be effective and efficient at attracting, screening, hiring, developing, and retaining top talent if analytics initiatives are going to deliver the desired target business value and economic impact. This is a time-consuming process that requires vision and the ability to sell people on the vision, getting the right people on the bus

DOI: 10.1201/9781003588344-12

and keeping them engaged, happy, satisfied, and productive. Hiring for "soft skills," such as communications, work ethic, attitude, and cultural fit, is as important as heavy-duty technical skills.

Many leaders, understandably but completely unrealistically, are looking for *"unicorns,"* i.e., someone who can do it all (Figure 11.1). You'll see what I mean if you take a look at pretty much any job description on any job posting for a data scientist. No individual is going to walk in the door and be an expert in all four required skill areas:

- Domain knowledge (i.e., industry, company, department)
- Data context and wrangling
- Mathematical and statistical modeling
- Technology (e.g., software engineering, programming, cloud computing)

Fortunately, however, these skills can be taught, but the bar is set depending upon the skill area.

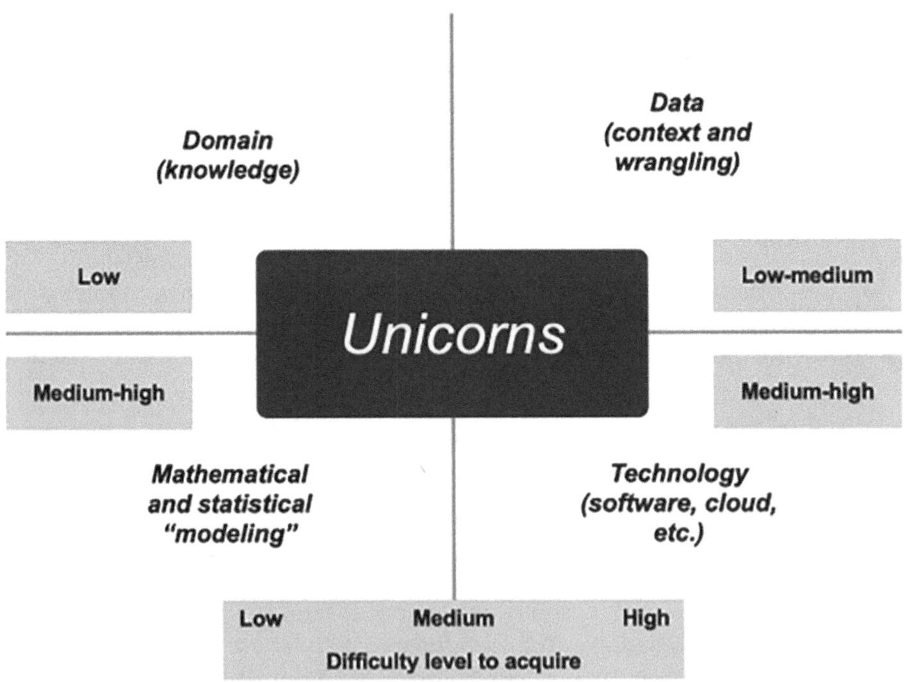

FIGURE 11.1
The four main required skills for data scientists.

Generating Demand (Securing Projects by Domain Area)

Doug had a boss (GM Jeff Honeycomb at McAfee) in the software industry, who used to say, *"Nothing happens until somebody sells something."* A primary role of the analytics/data science leader is to sell quantitatively oriented projects to the business people who need them. It's critical to identify projects that a) align with the company's strategy and performance targets (e.g., metrics, KPIs); b) can garner a high level of business executive/leadership support; c) secure the necessary funding and resources to execute the projects; and d) deliver the desired targeted business value and economic impact.

If you think of leading an analytical sciences team, no matter how large or small, it is like running your own business. You are, in effect, the CEO/COO *and* VP of Sales of that company. You will need strong delivery execution team members so you can spend the time getting to know the organization, people, and priorities, and hunt down and ferret out the projects with the most potential. Securing those projects will require trusted relationships with stakeholders.

Relationship Building

Trust arrives on foot and leaves on horseback. People don't care how much you know until they know how much you care. Trite cliché? Perhaps, but many analytical leaders believe that *trust* is critical to success, and I can state with certainty from my experience that is indeed the case. Analytics and data science often expose huge opportunities for performance improvement, which inevitably can make it look like someone isn't doing their job or is less than completely competent. Depending on the organization, it can take *years* of painstaking effort and a lot of coffees and lunches to build trusting relationships. A business leader has to have a *big problem* that they really need your help with before they risk their budget, career, reputation, and political capital on any project, let alone one involving a lot of math they don't fully understand.

Famed O.R. academic and practitioner, R.E.D. "Gene" Woolsey once said that *"A manager would rather live with a problem that they cannot solve, than implement a solution that they cannot understand."* Getting them to understand the solution requires a trusting relationship so they are comfortable that they will benefit from the project, and not get burned in the end.

Understand the Business Domain (and Problem in Question)

Gordon Bethune, the legendary former CEO of Continental Airlines, used to say, *"If you are going to run a watch company, you better make sure you know how the f@#$ing watch works."* In short, understanding the industry, corporate, and departmental business domain is critical before adding analytics or data science to it. There are often "business rules," e.g., contractual, legal, or regulatory implications, that constrain the decision or limit the full measure of benefit that can be achieved. As I first learned at American Airlines, executing analytics projects successfully requires far more than a rudimentary understanding of how an airline operates. And within an airline, applying analytics to revenue management or network planning is different from using them in network operations control or ground operations, or the more traditional business functions such as HR, finance, or marketing. You need to have innate intellectual curiosity and want to understand how things work before you can make them better with analytics. Everything starts with understanding the business problem; if you don't, your project is certain to fail regardless of how wonderful the math and code is. Period.

Change Management

Analytics and data science will often drive enormous changes in business policies, processes, and procedures; organizations; and jobs. Although analytics leaders may not need to be the "change management guru" in their companies, they need to be very sensitive to the shock waves that analytical results can have on an organization – the human impact as well as the financial, operational, and economic implications. They should also proactively engage and align with their change management specialists, if they are lucky enough to have them, to assist with the implementation of the new system and help ease the burden of the transitions. There will be inevitable changes in data requirements, systems, policies, processes, organizational structures, and decision-making (inserting the "model" in the human decision-making loop).

To be successful, the business must believe that the new and improved analytics-based process and system is *their idea* and in *their best interest*. And they should get to take all the credit for the value created.

My favorite quote on change comes from Niccolo Machiavelli in *The Prince* (1532). Even though he was talking about change in social and political systems, the principle applies to the change brought about by new systems based on analytical science.

"It must be remembered that there is nothing more difficult to plan, more doubtful of success, nor more dangerous to manage than the creation of a new system. For the initiator has the enmity of all those who would profit by the preservation of the old institution and merely lukewarm defenders in those who would gain by the new one."

When we look at the success rates of analytics projects, and IT projects for that matter, I believe it is clear that Machiavelli was onto something. Ignore the impacts of change caused by data science at your own peril.

Project Management

Analytical and data science models and systems should always be executed and delivered using a project management methodology approach, and in today's world, that is most commonly Agile (preferably Kanban for a modeling project, or Scrum for a project that is a part of a larger IT system development effort) or Scaled Agile (SAFe for extremely large, complex enterprise-level IT system projects). Scope, time, resources, and quality are the four primary dimensions of projects (Figure 11.2), and often scope (creep) is the most difficult to manage. Your goal should be to deliver more scope than you promise – easier said than done. These four project dimensions make up the "box" because no side of the square can lengthen or shorten without some of the other sides, or the enclosed project footprint *area* ($L \times W$) changing, which is not feasible. As an analytics leader, you will need to know how to wrestle with these factors, and how they interrelate, and there is as much, if not more so, art and nuance to it as there is science.

Most importantly, the project is not finished until you measure, and capture, the business benefit and communicate and act on the results.

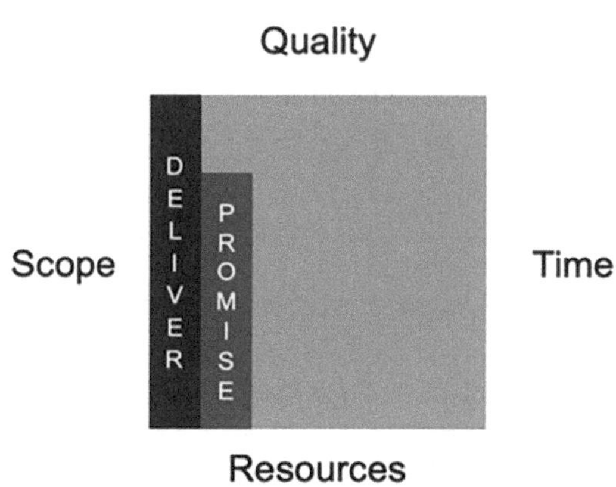

FIGURE 11.2
The four dimensions of project management.

Communication Skills

Even the greatest analytics or data science project using the most sophisticated techniques and delivering substantial business value and return on investment (ROI) may be rendered completely useless if you cannot convey its value and impact. Communications must be aimed at the appropriate audiences, including the business (at all levels, from the Board of Directors and C-suite down to rank-and-file individual contributors, and everyone in between) and technical (from PhDs to BS quant and non-quant grads) staff alike. Analytics leaders must develop the skills of how best to communicate complex concepts to each type of audience.

Some important communication pointers to utilize and master include:

- *Tell a story* that starts with a question or statement regarding a metric executives care about
  - "Using analytics, we were able to reduce inventory capital costs 25% year over year (YOY) with no adverse impact to availability, customer service, experience, and satisfaction"
  - This serves as a quick executive summary (start with the answer) to grab everyone's attention and keep their interest
  - Lead with the business value and economic impact; that's what executives care about
- Use a *"before and after"* rubric, i.e., what life was like *before* then *after* the analytics
  - Measure in metrics and key performance indicators (KPIs) that executives care about and understand
- Utilize more *visual images* (e.g., charts, graphs, heat maps, work flow diagrams, system schematics (stick to simple block diagrams)) and fewer words to tell the story
- Put all of the *math, code, and technical details* in the appendix

Planning, Budgeting, Administration, and P&L Management

Analytical and data science leaders should plan to run their teams like a business, i.e., your own company, complete with a profit & loss (P&L) statement and budgeting. Like any successful business enterprise, the economic value the group adds must offset its costs plus a sufficient return. Experience running either a consulting group or product-oriented business is a great training for this skill set. Gaining experience and developing substantial expertise in building and executing annual business plans, budgets, and proposals is imperative.

The *top line* (or sales revenue in most businesses) of your analytics business may be measured in incremental increased sales or revenue generated; customer lifetime value or market share; increased efficiency, throughput, or asset utilization; or reduced or avoided costs (operating or capital). The *costs* of your analytics business include labor, cloud computing and VPN costs, software license costs, data storage space, office space, etc. The *"bottom line"* or *"profit"* of your analytics business is the net of top line and costs. Over time, your organization will need to generate an ROI that *exceeds* the internal rate of return (IRR) that could be generated on *other* alternative projects using those same resources, i.e., opportunity cost. You will be evaluated and judged primarily on the tangible, measurable business value and economic impact that your team delivers YOY. Period. (Just like everyone else!)

Practitioner Experience and Expertise

Analytics and data science leaders, like many managers of knowledge workers, must be players/coaches, capable of at least occasionally working as a practitioner. Such an ability requires an in-depth knowledge of the analytical disciplines, i.e., methods, lexicon-jargon, KPIs/metrics, tools, and technologies. The best leaders of analytics organizations or functions will have worked *at least* 3–5 years as a "hands-on" analytics practitioner, i.e., one who builds models and analyzes data for a living as a primary job function. Understanding the underlying mathematical models, the data underlying the business domains, the technology underlying systems, and most importantly the *process and methodology* of developing, testing, verifying, and validating sophisticated models that solve complex business problems, requires hands-on experience to make judgment calls. *Try to imagine being the "chief of surgery" at a hospital without being a surgeon yourself, with all of the education, experience, and expertise that goes with that role.*

That said, a PhD is not necessary. I am a living proof of that rule. In fact, in my experience, what has worked extremely well for me is a:

- Quantitative BS degree (mine is in Math/Statistics, but also Engineering, Physics, etc.)
- Quantitative MS degree (mine is in Operations Research (O.R.), but also Statistics, Engineering, Computer Science, etc.)
    - A PhD is *not necessary*, but if you do have a PhD, a change in mindset from theory to practice, from research to applications, and from technical to business is *mandatory*

- MBA (concentrated ideally in Information Technology | Operations Management or Analytics, but doesn't matter; mine is in General Management)
  - Provides the deep skills required to run a business and manage a P&L

There are many more exceptions to the profiles mentioned above. I have seen folks with undergraduate degrees in everything from Philosophy to History to Business, who are self-taught and become superb practitioners and later leaders in analytical sciences. Tom Cook, who led AADT, had a BS in Mathematics, an MBA (from SMU), and a PhD in O.R. (from the University of Texas at Austin), so there you go!

Information Technology (IT) Experience and Expertise

Analytics and data science practice inherently involve *information technology* and lots of it. It is *impossible* to implement your model as a *microservice* or a *stand-alone system* integrated with other systems without understanding all of the layers and components of the tech "stack." Data pipelines, data warehouses, data lakes, and data lakehouses; homegrown and third-party enterprise software; computer languages, such as Python and Java; programming and software engineering; IoT, cloud computing using CPU/GPU servers, Agile for project management; and e-commerce and transactional systems all come into play. A laptop or JUPYTR notebook running Python alone will not cut it. Tom Davenport famously said, "*Models may make the enterprise smarter, but models embedded in production systems and business processes make the enterprise more efficient.*" Whether the analytics/data science team builds the systems in which to embed the models, or someone else does (perhaps the IT function), analytical leaders need to understand how to make models perform *at scale*, i.e., high availability, high reliability, and robust/fault tolerant, as a part of larger, more comprehensive enterprise systems and ecosystems.

My experience building model-based systems as project manager was foundational for developing the expertise and judgment required to lead and manage large, complex projects. Additionally, my CTO experience in leading the development of sophisticated e-commerce systems, like Travelocity and several others, as well as my VP of Engineering experience in leading the development of software products at McAfee was *crucial* for understanding the tech stack and commercial software product development processes, e.g., builds, QA, release train. My experience at Southwest Airlines as a director of Enterprise Data leading the development of enterprise-scale data warehouses and ETL data pipelines was invaluable to understand data engineering principles, processes, and best practices. I am not saying this is *required*, but any such experience would certainly be inordinately helpful in leading analytics programs and projects.

It's certainly true that sheer intellectual horsepower is always a useful trait to envision, design, and develop sophisticated, highly performant analytical models and solutions. Someone with a PhD in a scientific or quantitative discipline will no doubt have high levels of statistical and mathematical aptitude. However, I believe that pure IQ is less important than many of the other factors we have mentioned that are useful to effectively lead a high-functioning analytics team. Ample analytical abilities, along with emotional intelligence (EQ) (read *Emotional Intelligence* by Daniel Goleman) and other "soft skills" like those described above, are *critically important* to be an effective analytics leader. While my IQ is quite respectable, I was *never* the *smartest* person in any group I built, led, and managed; however, I was *extraordinarily adept* at the skills and capabilities listed in this chapter, and *that* is what enabled me to be a highly successful analytics leader.

12

Top 10 Reasons Analytical Sciences Projects Fail

Introduction

The failure rates of data science projects are well-documented. The number sits at about 80% of data science projects that never get implemented. The number also sits at about 80% of data science projects that fail to deliver business value. As a practitioner, these numbers bothered me because this was not my experience *at all.* I was fortunate to work for world-class analytics companies like American Airlines and Walmart, so I admit I was a bit biased.

This is a series of articles that began in 2020 as a PowerPoint presentation that I made to address *The Top 10 Reasons Why Data Science Projects Fail,* in response to the abysmal statistics about project outcomes. I presented it to the Data Science Community of Practice inside Walmart Global Tech and received a favorable response. Not long after, a colleague invited me to present on the topic in a session entitled Analytics Leadership at the 2022 INFORMS Business Analytics Conference. The meeting room at the Marriott in Houston was set up for about 75 people. About 100+ showed up, so it was *standing room only,* with folks standing in the back, sitting in the aisle, and lining the side walls. The response was overwhelmingly favorable, and many people came up to me afterward and said that I should write a blog to capture all of the great stories I told in my presentation. I then wrote about 50 pages to cover the topic, which was when I started thinking about writing a book but realized I didn't have enough content.

As luck would have it, I had another one of those career-defining moments. Kara Tucker, editor of *Analytics* magazine, published by INFORMS, reached out to me and said that she had heard about my presentation at the INFORMS Business Analytics Conference in Houston and asked if I would be interested in publishing an article on *The Top 10 Reasons Why Data Science Projects Fail.* I had found a home for the 50 pages, which was too long for one article, so she agreed to publish it as a *series* of 12 articles. As the series commenced, I posted the links to the articles on LinkedIn. *Herein lies the power of the global reach of social media.* A fine gentleman who I had never met or heard of reached out

DOI: 10.1201/9781003588344-13

to me via LinkedIn and asked if I would like to join forces, and content, to publish a *book* on this topic, as neither of us alone had enough material for a book. And the rest, as they say, is history.

Once again, the credit goes to Kara Tucker for introducing us to Randi Slack, a publisher at Taylor & Francis | CRC Press, where, as luck would have it, Kara used to work! Small world indeed!

Working together, Dr. Evan Shellshear and I researched, wrote, and published *Why Data Science Projects Fail: The Harsh Realities of Implementing AI and Analytics, without the Hype*, working closely alongside Randi Slack, our publisher, and Kara Tucker, our editor. For a *far, far more thorough treatment on this subject*, I highly recommend purchasing and reading our book. This chapter comprises the article series that started me down the road to really understanding in-depth *why data science projects fail.*

Why Data Science Projects Fail

Data science is by far one of the hottest technology domains and job markets ever witnessed on a global scale. Chief information officers (CIOs) surveyed by Gartner consistently rank data and analytics as one of their highest strategy and planning priorities. Gartner reported that the global analytics and business intelligence software market reached $21.6 billion in 2018 [1]. This market offers extraordinary business value and economic impact that corporations can realize by leveraging data about their customers, suppliers and internal operations, combined with advanced mathematics and (cloud) computing technology.

High-tech companies, such as Google, defined as "analytical competitors," use data science aggressively throughout their entire enterprise to sharpen operational performance and efficiency and improve customer experience in their retail and online search businesses, respectively. Companies like American Airlines pioneered the use of data and analytics in the field of revenue (yield) management in the 1980s to generate $400–$500 million in incremental revenue annually. UPS saves $300–$400 million annually with its On-Road Integrated Optimization and Navigation (ORION) application that guides their 55,000 delivery truck drivers every day. Walmart generates millions of dollars in value annually by applying predictive and prescriptive analytics to optimize its markdown pricing strategy. (The project is actually a 2023 Franz Edelman Award finalist.)

Despite these genuine success stories and many, many more examples, according to a study by Deloitte Analytics and Tom Davenport, only 20% of data science models built are actually deployed into a production system supporting a business process. Gartner reported that through 2022, only 20% of analytic insights will deliver business outcomes. This means that organizations are investing billions of dollars in analytics with minimal return – hardly a recipe for success [1].

Why do most data science projects fail to get deployed and deliver the desired business value, outcome or economic impact? A series of monthly articles will fundamentally focus on this question. We will explore and explain the organizational and individual behaviors and factors that contribute to most data science project failures, which must be addressed and consciously practiced to increase the likelihood of success. Spoiler alert: The problem is not with the mathematics and technology but rather with the actions of the people (practitioners and leaders) engaged in and the processes employed to execute and manage data science projects.

As a long-time proponent of Stephen Covey's "The 7 Habits of Highly Effective People" [2], I find two of his habits particularly useful in the endeavor to create more successful data science projects:

1. **Begin with the end in mind**. Analytics expert, researcher and author Tom Davenport said, "Models make the enterprise smarter; models embedded in systems and business processes make the enterprise more economically efficient." This should be your end goal when starting work on a data science project. You don't want to just build a model; rather, you want to embed that model into a mission-critical system that supports a key business process such that greater economic efficiency (i.e., lower cost, greater revenue, improved customer experience) can be achieved on an ongoing basis in an automated manner with little or no human intervention, creating a flywheel effect generating business value.
2. **Sharpen the saw.** Abraham Lincoln once said, "If I had six hours to cut down a tree, I'd spend the first four sharpening the saw." Most undergraduate and postgraduate education program coursework in data science (and related fields) is spent focused on mathematics and computer science methods, skills, and technologies. Although this is understandable because (1) considerable training in these domains is necessary to become a data science practitioner and (2) this is what university staff know how to teach, students enter the workforce unaware of the more nuanced, subtle and harder-to-grasp aspects and dimensions of executing, managing and leading data science projects in the real world and corporate America. My intent here is to help students, practitioners, leaders and executives "sharpen the saw" and fill in the knowledge gap in their training and education that heretofore was learned only through real-world work experience.

This article series is a compendium of my own experiences, and observations of other practitioners, from more than 30 years as a practitioner, leader and executive in corporate operations research, analytics, data science, data and software engineering, e-commerce, and consulting organizations and as a business analytics and data science educator/researcher at Southern Methodist University's Cox School of Business and other continuing and professional education programs. Given the low success rates of data science projects to deploy and deliver value, I felt compelled to share what I and others have learned with the goal of helping practitioners and leaders be more successful more frequently and avoid many of the common pitfalls associated with these endeavors.

Originally, I presented this material to the Data Science Community of Practice inside Walmart Global Tech in 2021, and then again in April 2022 in Houston, Texas, at an INFORMS Business Analytics Conference in the Leadership Track under the title "The Top 10 Reasons Data Science Projects Fail." About 100 people attended my talk, in a room with a capacity of about 70, and the feedback was so overwhelmingly positive that several people said to me, "You should really write all this information down!" That particular invited speaker address was the genesis of this series!

The objective of the material in the series is to help make you a more well-rounded, self-aware, and informed data science practitioner and leader by learning from the experiences gained by others in the field who came before in the spirit of "fail fast and learn."

Although I utilize "data science" as a contextual delimiter, the series' principles apply equally to related adjacent fields that utilize data and mathematics to model and solve business problems and phenomena, such as operations research/management science, statistics, analytics, machine learning, artificial intelligence (AI), business intelligence (BI) and more.

Samuel Smiles famously said that we learn more from failure than we do from success. My approach therefore was to examine some of the primary reasons I have observed data science projects fail – a top 10 list, if you will – to highlight to data science practitioners the aspects and dimensions of their projects that are more subtle, less tangible, and more difficult to grasp, but no less critical, and of which they need to be more conscious to generate successful outcomes with greater consistency and regularity.

"We learn wisdom from failure much more than from success. We often discover what will do, by finding out what will not do; and probably he who never made a mistake never made a discovery." – Samuel Smiles

The series will comprise 10 articles, each tackling a different reason why data science projects fail and how to address them.

Now, join me on the journey to find out why data science projects fail and learn how to avoid making the same types of mistakes.

References

1. K. Troyanos, 2020, "Use Data to Answer Your Key Business Questions," *Harvard Business Review,* February 24, https://hbr.org/2020/02/use-data-to-answer-your-key-business-questions.
2. Stephen Covey, 1989, *"The 7 Habits of Highly Effective People,"* New York: Free Press.

Part 1: What's the Problem (That You're Trying to Solve)?

> If you try to tell someone else how to do their job better using sophisticated mathematics and computers without thoroughly understanding how they do their job today, including all of the problems and challenges they encounter, then you sir/madam are a fraud.
>
> – *R.E.D. "Gene" Woolsey, Ph.D.*
>
> *Professor, Colorado School of Mines*
>
> *Operations research academic, practitioner and consultant*

Businesspeople and data scientists, individually and collectively, not understanding the real business problem at hand is the no. 1 reason data science projects fail, in my experience. Most often, a data scientist will collect data and build a model to only, at best, come up with the right answer to the wrong problem – i.e., a problem or question that the customer did not convey. Communication is a big issue that we will talk about later, and that is part of the challenge here, but there are several foundational steps that a data scientist must take before engaging on a project to help ensure that the real business problem that is being addressed is mutually and thoroughly understood.

First and foremost, data science fundamentally requires a high degree of intellectual curiosity to be done well. You cannot be a data scientist "at arm's length." You will need to take a deep dive into and "get dirty" with the details of the company's industry and business. To be effective, a data scientist requires deep contextual understanding at three levels:

- Industry and segment.
- Corporation and department.
- Domain problem space.

Data science applications vary greatly across industries and their respective segments from energy (oil and gas, electric, wind, solar, generation, transmission) to transportation (airlines, railroads, trucking, rental cars), healthcare (providers, insurers, device manufacturers, pharmaceuticals), financial services (banking, credit cards, credit reporting, mutual funds, hedge funds, private equity, venture capital), manufacturing (automobiles, steel, consumer packaged goods, semiconductors, food) and retail (big box, hardware, clothing, housewares). Each of these industries has their own unique economics, operating models, and competitive landscapes. It necessarily behooves the data scientist to research and understand as much as possible about the industry in which one is working.

Each corporation within a given industry or segment has its own competitive and economic DNA (e.g., low-cost provider versus premium high-margin provider), culture, and mode of operation. A data scientist must learn and understand the following:

- What is the company's business model?
- What is the company's strategic competitive advantage?
- What is the company's core product and/or service offering(s)?
- How does the company make (or lose) money (e.g., order-to-cash cycle)?
- What are the primary sources of sales, revenue, and cost (operating and capital expenditure)?
- What makes the company "tick"?

The company's *annual report* and *financial statements* are a great source of in-depth, detailed information to learn about the above topics. (If you don't have a BBA/ MBA/CPA, then find a friend in accounting or finance to help you get started!)

Inside your company, many, many different departments may be using data science (or none, depending on the company's data and analytical maturity). The approach to data science and the problems to be solved are as varied as the department:

- Marketing
- Sales
- Manufacturing
- Operations
- Finance
- Accounting
- HR

You will need to understand the goals, objectives, business processes, metrics, operating plans, and road maps of the organization with which you are working to apply data science. You need to know *how the work gets done,* including budgets, data, data systems, and software. You literally need to learn to speak their language – and, yes, each department *will* have their own vocabulary, terminology, and acronyms (corporate America *loves* acronyms). Being a data scientist requires deep immersion in your industry, your company, and your department to understand the domain problem space and be able to contribute materially – your goal is not to appear to be the "math geek with the fancy laptop" but rather to be a "team member that digs deep and helps solve problems using some really powerful, specialized skills."

When I started working for the American Airlines Operations Research Department in 1987, fresh out of graduate school at Georgia Tech, all I knew about airlines was making a flight reservation, getting a boarding pass, finding my seat, ordering a drink, and claiming my luggage. Over the subsequent six years, I learned about all of the relevant facets of airport operations, airline operations, maintenance/inventory operations, and crew/flight academy operations. Whenever I had a new project, I *physically parked myself* in the problem area next to the people who did the actual work – i.e., the air traffic control tower and radar approach control center, network operations control center, maintenance hangar, and office building – and I didn't leave until I understood how they did their job and the current problem that we were trying to solve. Then, and only then, did I commence with the data science modeling work.

There are several questions data scientists need answers to before beginning a project:

- What is the problem that we are trying to solve, clearly and succinctly stated?
- What is the key business question we are trying to answer?
- What is the desired business outcome?
- What is the end state of the model/system we build? How will it be utilized?
- What is the "target" for improvement (e.g., cost reduction and conversion rate increase)?
- What KPIs (key performance indicators) are relevant to measuring economic impact?
- What experiments can we run to measure the before-and-after effect of the model?

Multiple meetings and whiteboard sessions may be required to adequately answer these questions, but it will be time well spent for all parties involved. As the old software engineering adage goes, "An ounce of design is worth a pound of debugging."

Gartner reported that "through 2022 only 20% of analytic insights will deliver business outcomes" [1]. Therefore, understanding the true business problem at hand will help your project make the 20% cut line!

Reference

1. K. Troyanos, 2020, "Use Data to Answer Your Key Business Questions," *Harvard Business Review*, February 24, https://hbr.org/2020/02/use-data-to-answer-your-key-business-questions.

Part 2: Data, Data Everywhere ... But Not in One Location to Analyze

> We don't have the data!
>
> – Said almost every client with whom I have consulted, and many students I have taught, when confronted with the prospect of doing a real-word data science project.
>
> If your data is everywhere, then it is nowhere.

Data issues are typically the No. 2 reason data science projects fail. Data issues manifest themselves in myriad ways as varied as there are companies attempting to manage and analyze their data. (See Tom Davenport's book, "Big Data at Work," for a great resource in this domain.) (https://www.amazon.com/Big-Data-Work-Dispelling-Opportunities/dp/1422168166)

The first complaint I usually hear from clients asking me how to do analytics in their enterprise, and from students when confronted with the reality of having to do a real-world data science project in one of my courses, is, "We don't have the data!"

That may be true because many companies, especially small to medium-sized businesses, are bereft of (automated) data altogether or lacking clean, accurate, consistent, high-quality, and high-integrity data.

More often, what people really mean is that the data is not all in one place – the most common data affliction of most enterprises. Data is literally scattered among dozens, or even hundreds (no exaggeration), of enterprise applications, legacy systems, databases, CSV files, data warehouses, data marts, cloud accounts, third-party systems and, yes, the ubiquitous Excel spreadsheets.

A data scientist alone, in most enterprise instances, is not going to be able to solve this problem in isolation. They need to partner up with IT, a database administrator, a data architect, cloud data engineer, or the chief data officer's data engineering team (if you are lucky enough to have one).

Although you should not try to "boil the ocean" (because you will figuratively "drown") and solve *all* of the enterprise's data issues, you should stay laser-focused on organizing the data you need for your project. That will be challenging enough.

Two main issues to focus on to enable your data science project are (1) data integration and (2) data governance.

Historically, the database, data warehouse and data mart were the common enterprise data stores. Recently, these have been superseded by the data lake (usually unstructured, raw data landing zones) and now, yes, the data *lake house*, which combines attributes of the warehouse and lake into one entity. Regardless of the exact data platform, getting all of your data cleaned, organized, and integrated into a single physical or virtual (accessible) workspace or view is critical to enabling your data science project out of the gate.

Some companies, such as large retailers, airlines, telcos, and financial services, are blessed, and cursed, with enormous amounts of data, i.e., duplicates and sheer voluminous amounts of data. This is a good problem to have from a data richness perspective but can represent logistical problems of storage and management.

Data governance, including metadata, data lineage, and data stewards, is a hot topic and an absolute necessity to ensure one version of the truth, consistent data definitions, and usage patterns. Once again, the data scientist will not solve this problem alone either but will need to partner with the data governance team (if you are lucky enough to have one) or at least data owners and stewards that control access to and governance of some or all enterprise data.

Personally, I have been greatly blessed when it comes to data. I worked for companies that were rich in data resources, relatively mature in the way in which data were managed, and legitimately data-driven and analytically inclined, including American Airlines, Sabre, Southwest Airlines, Blue Cross Blue Shield of Kansas City, and, most recently, Walmart.

At Southwest Airlines, I was in charge of the enterprise data warehouse Teradata and ETL & Reporting (as well as advanced analytics), and we delivered some very substantial, very challenging projects, such as the Reservation Data Pipeline & Warehouse (3,900+ test cases) and Customer Data Warehouse. We created whole new data structures to support a brand new jet fuel demand forecasting, purchasing, and inventory management, replacing 150–tab spreadsheets and pulling data from a half-dozen legacy and new transaction and information systems. That effort avoided millions of dollars annually in superfluous jet fuel costs using advanced analytics. But without integrated and governed data and automation, data science is just math!

Part 3: Misapplying the Model

> There are many ways to do something wrong, but only one way to do something right.

Sometimes data scientists, especially ones with insufficient education, training and practical experience, make the mistake of incorrectly applying a model to a problem.

Let's start with a real-world example that I personally encountered.

An airline website merchandising manager wanted to test which of two landing (web) page designs would capture the most customers interested in purchasing air travel on a commercial US airline. This is a classic problem and properly calls for A/B testing, which anyone can run as an experiment using Google Analytics.

Prospective customers are randomly shown one of two landing page designs (A or B) and choose whether to click through to the airline travel offer content. The results are tallied as a ratio of the number of those customers who clicked through landing page A divided by the number of total customers who were randomly shown page A, and similarly for page B. The ratio measures the success rate, or "hit rate," for each landing page:

$$\text{Hit rate}_A = \frac{\text{Number of customers who clicked through page A}}{\text{Total number of customers shown page A}}$$

$$\text{Hit rate}_B = \frac{\text{Number of customers who clicked through page B}}{\text{Total number of customers shown page B}}$$

A/B testing is a standard and appropriate use of Pearson's chi-square test for independence to determine whether there is a *statistically significant difference* between the two ratios, such that if a material difference did exist, it would indicate that one page is more effective than the other at attracting customers to view the offer.

The *misapplication* of A/B testing manifested when the manager wanted to use the *exact same test* to determine which page would generate more *revenue.* Whereas hit rate is a simple ratio of page click-throughs to total page views, revenue is a far more complex, multivariate, and multidimensional quantity. The myriad variables that determine the total revenue on an airline website transaction include, but are not limited to, the number of ticketed passengers on the itinerary, origin-destination market pair (e.g., DAL-PIT and DAL-LGA), fare class, fare class bucket, purchase date relative to flight date and so on. You get the picture.

To determine with statistical accuracy which landing page generates more revenue, you would need to rigorously design a series of highly controlled experiments to account for most, if not all, of these variables that materially affect revenue to ensure that you are comparing "apples to apples" and not "apples to oranges."

In my experience, as a practicing statistician applying my education and training, I find that a fundamental lack of understanding by professional and citizen data scientists of the *principles of experimental design,* such as in the airline web page example, is one of the biggest "gaps" in solving these types of real-world problems (i.e., comparing two alternatives against a metric). We work in a world that is complex and multivariate with confounding effects, and we must account for all of that in our data science projects. Entire books have been written on the topic of *experimental design,* and the practical applications of these techniques in industries as different as farming to pharmaceuticals all share the goal of legitimate, logically and statistically accurate results, conclusions and decision-making.

A classic example of experimental design is the testing of two fertilizers (A and B) to see which one generates greater crop yields. If you had a field next to a river running rich with nutrients, you would not want to plant crops using fertilizer A alongside the river and fertilizer B further inland. The *effect* of the river would *confound* the experiment by giving an unfair advantage to fertilizer A – more nutrient-rich soil. You need to *control* the effect of the river to ensure that both fertilizers have equal access to river-enriched soil. Therefore, you would plant the two fertilized crops *adjacent* and *perpendicular* to the river. (See Figure 12.1 for an illustration of the incorrect and correct fertilizer experimental designs.)

Alternatively, an experiment to test for the efficacy of a new pharmaceutical drug, such as a blood thinner, would need to control for a variety of variables, including (perhaps) gender, age, weight/body mass index, cardiac disease history, blood pressure, pulse rate, overall health, and genetic makeup, among many others.

The point here is that for as many models that exist, there are many, many ways to misapply them. Experimental design is one of the easiest and most common traps into which a data scientist or businessperson can fall. Overfitting and bias are also model conditions to aggressively guard against

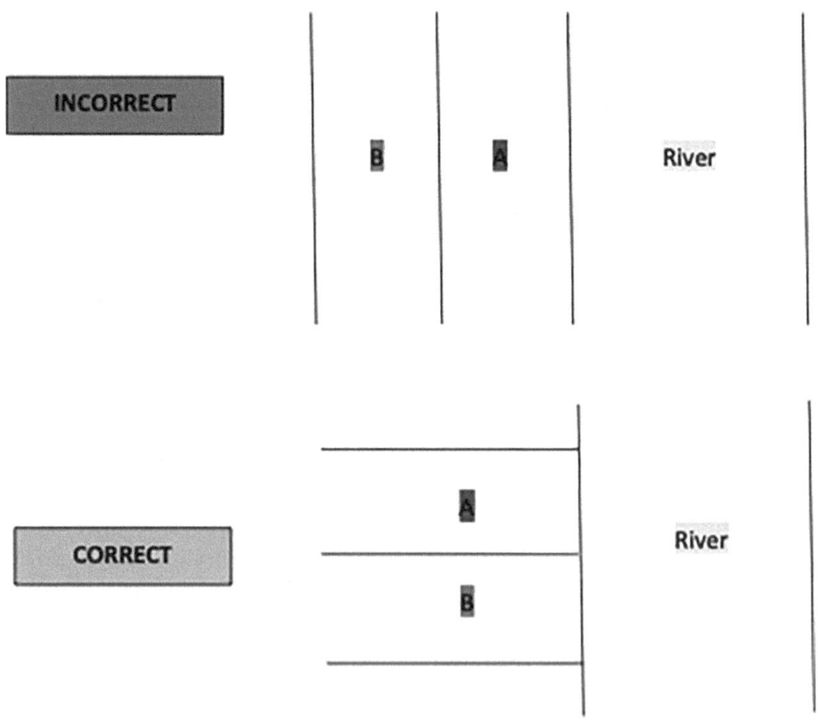

FIGURE 12.1
The concept of experimental design, visualized using the example of testing two fertilizers.

to avoid erroneous results in predictive analytics modeling scenarios. One must take great care to consider the problem context and proper fit of the model's application with the problem at hand.

Sometimes, there is more than one model form that may be legitimately applicable to a given problem context. For example, to predict the likelihood of the presence of cancer in a patient on the basis of biopsy and X-ray results data, data scientists might use logistic regression, Bayesian inference and/or an artificial neural network, but they would not use linear regression. With today's available cloud computing power, one can easily fit all three models against the data and compare how they perform using a *confusion matrix* to determine which is best.

At the time of model selection – early in the project – it is always a good idea to consult with a colleague, either a peer or former professor, as a sounding board on the best approach. The art of understanding the best model and experimental design for the problem at hand is as important, if not more so, than the science. The old carpenter adage "measure twice, cut once" is apropos here.

Part 4: Solving a Problem that Is *Not* a Business Priority

> Ruthlessly rigorous prioritization of (technology, data science) projects based on potential business value and economic impact is the best way to ensure meaningful, successful outcomes.
>
> – *Fortune 5 EVP|CTO*

Every company has limited capital and human resources in IT and data science. There are *never* sufficient budget dollars and people to go around to fund all projects. In cases I've observed in Fortune 50 companies, new project demand exceeds the available budget by a factor of two to four times. Projects must compete for resources during each budget cycle based on their respective relative potential to generate incremental business value.

Data science projects are no exception and are ultimately judged on their ability to "move the needle" on economic performance. That said, in many companies, a lot of political wrangling and "pet project" machinations go into the decisions as to which projects get worked on, i.e., the **HiPPO** projects – **H**ighest **P**aid **P**erson's **O**pinion.

Fortunately, there are multiple rational, fact-based, data-driven frameworks that help estimate, gauge and compare value and inform data science project decision-making.

In a *Harvard Business Review* (HBR) article by Kevin Troyanos, "Use Data To Answer Your Key Business Questions," a heuristic rubric is offered to help prioritize business questions using a two-dimensional grid. (See Figure 12.2 for an illustration of The Key Business Question Grid.)

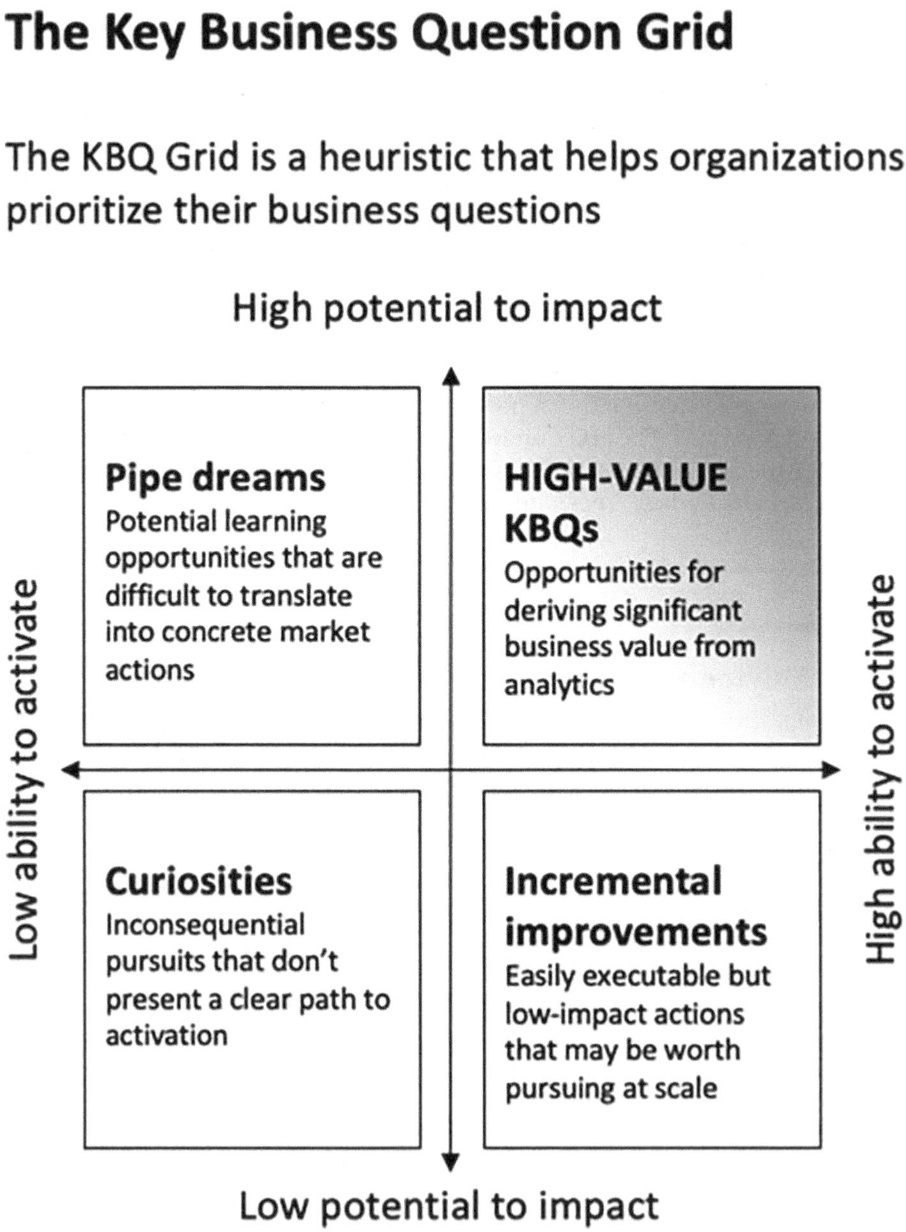

FIGURE 12.2
The Key Business Question (KBQ) grid.

- *x*-axis (horizontal) – Ability to activate or ability to execute, implement, and deploy the solution.
- *y*-axis (vertical) – Potential to impact business or economic value potential.

High-value key business question (KBQ) projects in the upper right-hand quadrant of the HBR figure, with high ability to activate and high potential to impact, are what you and your business partners want to be working on most of the time. Selecting projects that can be implemented and also deliver significant, tangible, measurable business value and economic impact (e.g., cost reduction, operational efficiency or performance improvement, revenue increase, or customer satisfaction and experience enhancement) can be challenging but is absolutely necessary for long-term success.

Curiosities should be completely avoided, as they consume resources on projects that offer low ability to activate and low potential to impact.

Pipe dreams, sometimes referred to as "moon shots," offer high potential to impact but a low ability activate. Sometimes, companies embark on such projects, despite a low likelihood of success, on the fervent hope that they will succeed and deliver tremendous market leverage or competitive advantage. If the project fails, then they try to extract key learnings to feed into other less ambitious projects.

Incremental improvements offer low potential to impact and a high ability to activate. These types of projects can be effective when an organization is just getting started with data science and is looking for some "quick wins" while they are building momentum and growing the capability to handle larger, more complex and higher-value initiatives. The payoffs are not as great, but if they add some measurable value, and the team sharpens their skills, then that is a win.

Because there will always be many views, perspectives and opinions on how best to prioritize and select among competing projects, a more uniformly applied, objective approach can help level the playing field. For those who prefer a more *quantitative* approach to *scoring* and *ranking* analytics, data science and artificial intelligence (ADSAI) projects, I have successfully utilized the following process in a variety of software and analytics product/solution development settings. (You can easily do this in Excel – one of the only times I will recommend a data scientist use Excel!) See Figure 12.3 for the Excel-based ADSAI project scoring and ranking spreadsheet rubric generated from the process below.

Projects	Bus Value$	BusValScore	Complexity	CmplxScore	Resources	Res Score	Score
Project 8	$11,000,000	6	LOW	8	$ 2,500,000	5	240
Project 2	$ 8,000,000	4	LOW	7	$ 1,800,000	7	196
Project 9	$ 5,000,000	2	VERY LOW	10	$ 1,000,000	9	180
Project 5	$ 7,000,000	3	LOW	7	$ 1,750,000	8	168
Project 4	$12,000,000	7	MEDIUM	5	$ 3,500,000	4	140
Project 1	$10,000,000	5	MEDIUM	4	$ 2,000,000	6	120
Project 10	$ 4,000,000	1	VERY LOW	10	$ 500,000	10	100
Project 6	$14,000,000	8	HIGH	3	$ 4,000,000	3	72
Project 7	$16,000,000	9	VERY HIGH	2	$ 4,500,000	2	36
Project 3	$20,000,000	10	VERY HIGH	1	$ 5,000,000	1	10

FIGURE 12.3
A quantitative approach to prioritizing data science projects.

1. Start with a list of your projects by name in Column A, one project per row.
2. Estimate the *business value potential* in dollars that each project will generate in Column B.
   a. For each project, on a scale of 1–10, where 10 is highest business value potential, put a business value potential score in Column C.
3. Assess each project's *complexity* as VERY HIGH, HIGH, MEDIUM, LOW or VERY LOW in Column D (if it helps, think as if you were doing Planning Poker estimating Story Difficulty Level with Fibonacci numbers in Agile Scrum).
   a. For each project, on a scale of 1–10, where 10 is lowest complexity, put a business value potential score in Column E.
4. Estimate each project's total resources – i.e., labor time, materials, computing – in dollars in Column F.
   a. For each project, on a scale of 1–10, where 10 is lowest cost, put a total resources score in Column G.
5. Multiply the scores for each project in Columns C, E, and G and put the resulting product in Column H.
   a. The *maximum* score is 10*10*10=1,000 (which would indicate a project with highest relative business value potential, lowest relative complexity, and lowest relative cost).
   b. The *minimum* score is 1*1*1=1 (which would indicate a project with lowest relative business value potential, highest relative complexity, and highest relative cost).
   c. Each project now has a score from 1 to 1,000.
   d. You can think of these scores as a surrogate quantification of the KBQ grid (from the HBR article) for each project's potential to impact (value and cost) and ability to execute (complexity).
6. Sort the project rows from highest to lowest (score) on Column H.
7. The result is a prioritized list of data science projects, using an objective, quantitative scoring mechanism.

The *primary takeaway* from this exercise is that the *multiplicative scoring* approach ensures that not just high business value projects bubble to the top of the list, rather, the magnitude of business value is tempered by a combined effect of complexity and cost. *Complexity*, in effect, is an important surrogate measure for *risk*, i.e., the more complex a project is, the more likely you are to run into difficulties that end up manifesting themselves in timeline delays and budget overruns that jeopardize the whole project. In the chart, we see the *highest scoring projects* are those that have low-to-medium relative business value moderated by (very) low complexity and low-to-moderate costs.

The highest-value projects, in this example, happen to have the highest risks and costs, which result in a lower score. This is actually a fairly commonly encountered set of circumstances, i.e., high risk/cost, high reward.

Now is when the "fun" begins and people start debating and haggling (i.e., arguing) over the individual and aggregate score for their respective project(s). (This process should be accompanied by a more rigorous financial analysis using NPV, ROI, and internal rate of return metrics.)

This is not just a theoretical exercise. When I was a VP of Engineering & Product Management for a division of a $1.3 billion software product company, we had a list of 2,000 feature modification requests (FMRs). We estimated that we had capacity to do about 500 FMRs in a new major product release. We used the above process to rationally, objectively, and as economically and efficiently as possible narrow down the list of projects our engineering team could realistically do in one release cycle. It really does work in practice!

Part 5: Effective Communication

> What we have here is a failure to communicate.
>
> *– Captain (played by Strother Martin) in "Cool Hand Luke"*

Clear, concise communication – verbal, written, nonverbal – or lack thereof represents a significant challenge in business in general upon which we all strive to improve. In the field of data science, the communication challenge is even more acute for several reasons, not the least of which is that business people and data scientists rarely speak the same language. Data scientists, who speak a language of mathematical models and symbols and "code," must endeavor to understand the business domain and problem space that is defined by terminology and acronyms that are inherently foreign. Managers who speak a language of KPIs and business jargon and acronyms must try to understand how a complex, sophisticated mathematical model, automated on a computer, is going to solve their business problem. Data scientists must be intellectually curious and dig deep to understand the business problem and context. Managers, who are likely not mathematicians themselves, must "trust but verify" through rigorous experimentation, verification, and validation that the model and solution (inside the "black box") are functioning appropriately to solve the problem at hand.

It is human nature that people inherently abhor change and fear what they cannot understand. Famed operations researcher Gene Woolsey said, *"A manager would rather live with a problem they cannot solve than accept a solution they cannot understand."* The communication challenge then requires the data scientist and manager to create a conceptual and practical intersection between their two worlds in which they can communicate and understand each other.

One of my favorites of Stephen Covey's "Habits of Highly Effective People" in the communication realm is #5: *First seek to understand, then be understood.* This habit compels us to listen before we speak. The old adage that we have two ears and one mouth so that we listen twice as much as we speak provides an excellent heuristic for data scientists to govern their communication approach. Throughout a project, but especially in the early stages, data scientists should be listening *two-thirds* of the time and speaking *one-third* of the time, and when they are speaking, they should be *asking exploratory and clarifying questions*. This ratio will tend to change toward the end of the project to be more 50–50 listening and speaking as the data scientist explains how the model and system works; presents findings, conclusions, and recommendations; and answers what will no doubt be a myriad of questions from the manager.

In every data science project, it is critical to consider the context and audience in each situation and adapt your communication approach and content accordingly. Are you, the data scientist, talking to another data scientist on your team or on the business side? Are you talking to the manager on the business side – or perhaps their up-line executive, such as a director, vice president, or above – perhaps giving a demonstration or presentation of your model and solution? Maybe you are talking to a business analyst equipped with an engineering degree or quantitative MBA who has a much better understanding of data science models, computers, and software applications than their manager. Know your audience, prepare, and communicate accordingly.

Communication is crucial and should be engaged in before, during, and after the project.

- *Before* the project to mutually set expectations on scope, timing, budget, critical success factors and criteria, and to achieve a crystal-like mutual clarity of the problem at hand and the solution approach.
- *During* the project to ensure tight feedback loops because modeling is by nature and necessity *iterative*, and not necessarily strictly "linear," and to provide updates on status and negotiate changes in direction or approach as new information and discoveries come to light.
- *After* the project to communicate and act on the findings, results, conclusions and recommendations and most importantly, to quantify the business value and economic impact of the model.

After the acceptance of the model and solution comes the substantive communication that must go into implementing the model as part of the business process (the next installment will cover this and change management).

One piece of advice on communication media: *Avoid email if at all possible,* especially on critical, sensitive topics. Email is a horrible communication vehicle for nuanced, complex information sharing. Data science project communication is the utmost in nuanced, complex information sharing, and email creates myriad opportunities for misinterpretation and misunderstanding. There is no substitute for face-to-face communication, whenever possible, even via video conference in the new post-COVID-19 age of remote work.

Stories as Communication

The most effective data scientists are *storytellers.* They tell a story of what life was like before the model was developed and implemented and how life will change (hopefully for the better) afterward. They start presentations by grabbing the attention of the audience – in particular, executives who are prone to reading the news, email, or their calendar on their mobile devices. The most effective data scientists ask provocative rhetorical questions such as, *"What if I told you that we could increase sales (or decrease inventory costs) and make (or save) the company an extra $X gazillion using data and data science?"* Now you have everyone's attention! The *key* is to communicate in the language of your audience – i.e., managers, executives, and domain experts – not data science!

Lastly, for any data science project to move forward, you will inevitably have to address and adequately answer the age-old question: *"What's in it for me, my team, my department, the company?"* As the late NBC Sports television executive Don Ohlmeyer once said, "The answer to all of your questions is money." The answer may be operating or capital expenditure cost savings or avoidance, increased revenue, increased customer satisfaction, or increased resource utilization, all of which may lead to some economic improvement for the people involved (like a bonus, raise, or promotion!) or the company at large (higher stock price, increased dividend, increased profit sharing, etc.). Everyone wants to understand how they, and their stakeholders and constituents, are going to benefit by undergoing this cataclysmic change in their business process.

Part 6: Change Management

> There is nothing permanent except change. All is flux, nothing stays still.
>
> – *Heraclitus*

> It is not the strongest or most intelligent who will survive, but those who can best manage change.
>
> – *Charles Darwin*

Despite the unassailable veracity of these age-old adages, most people, in general, don't appear to be getting any more adept at or comfortable with accepting, embracing or effectively dealing with change.

Technology, in particular over the past 50 years, has dramatically accelerated the pace of change in business. From robots building cars to meetings being conducted via video conference to AI-based systems streamlining, automating, and optimizing large-scale complex decision-making and problem-solving better, faster, and more effectively than any human ever could, change is daunting for most people. The fear that jobs and livelihoods will be eliminated can be damaging to the psyche.

The simple fact is that data science is disruptive. The sheer volume of data and the dynamism and complexity of business decision-making and problem-solving mandate the use of automation and mathematical logic and intelligence. However, when we "let the data speak," inconvenient truths are revealed. Gut instinct and heuristic "rule of thumb" planning, decision-making, and problem-solving processes that sufficed for decades are now invalidated and outmoded by new and improved fact-based, data-driven, and model-based solutions. Solution-embedded models, be it in a robot or a business system, radically change the way we work, make decisions, and solve problems. This in no uncertain terms *threatens* the human beings who are used to doing things "their way, the way they have always done it for 30 years with their 150-tab Excel spreadsheet" (no exaggeration, this is an actual fact from a project I worked on).

Change management is without a doubt, one of the most critical dimensions of successfully executing a data science project. The data scientist must gently, diplomatically, and ever so delicately win the hearts and minds of the businesspeople who will be instrumental in designing, developing, testing, validating, approving, deploying and ultimately using the new system. If you are unable to convince and bring those folks along the journey, then you will lose their support, and your project will fail with 100% certainty. Period.

It is not enough to *tell them* that your "black box, math-magical computer system" is better, faster and more economical than "the old way." You need to *show them*, step by step, from beginning to end and case by case by making them an *integral part of the process* and not the *recipient* – or worse, the *victim* – of it. Change can be a painstakingly slow process as businesspeople move through the full range of emotions and reactions to the new solution, including (not dissimilarly to the five stages of grief) denial, anger, bargaining, depression, and acceptance (hopefully). You can throw in outright rejection of the new solution in favor of the incumbent:

> It must be remembered that there is nothing more difficult to plan, more doubtful of success, nor more dangerous to manage than the creation of a new system. For the initiator has the enmity of all those who would profit by the preservation of the old institution and merely lukewarm defenders in those who would gain by the new one.
>
> – *N. Machiavelli [1]*

Principles of Change Management

The first principle in change management is *empathy*. Be kind to the businesspeople who are your stakeholders and constituents. Your tone, body language, mannerisms, and communication style all need to soften the sharp edges of the pure rationality and logic of the data, math, and code. Put yourself in their position and look at the situation from their perspective. This goes back to Part 5 on Effective Communication – *"first seek to understand, then be understood."* Position and present yourself as a colleague on the same team, not the enemy trying to disrupt their processes. Data science is intended to make things better, economically for all parties, not just be more change for change's sake.

It is important to remind folks that data science, artificial intelligence, and machine learning (AI/ML) are more about *augmentation* than *replacement* (for the foreseeable future, anyway). The "human-in-the-loop" working interactively and iteratively with the model, not being replaced by it [2].

I learned all about change management firsthand in 1990 on my first project to build a new decision support system from scratch for American Airlines to schedule aircraft heavy maintenance checks and plan hangar capacity for a five-year planning window. Historically, the airline with 200 aircraft created their long-range five-year heavy maintenance and hangar plan on large sheets of paper using colored pencils, driven by calculator computations of when aircraft would be due for their respective checks. When the fleet rapidly grew to 600 aircraft, the paper-pencil-calculator solution became unwieldy and untenable, so the analysts (two at the time, later three total) switched to Excel macros. Unfortunately, the macros sometimes took as long as ten hours to run to completion, for the larger subfleets, on an Apple Macintosh IIcx desktop computer (Motorola 68000 chipset), and many times, they errored out prior to completion. Senior management quickly grew impatient and very, very concerned as the mystery loomed of when a new hangar might need to be built, and check yields were bleeding down to a suboptimal 80%, increasing heavy maintenance costs [3].

An industrial engineer at the maintenance base had done an analysis that demonstrated whether a system could be built to *automate* the maintenance check and hangar planning and scheduling process and *optimize* the schedule to maximize the check yields to ~100%. (Two heavy maintenance checks, $1 million each or $2.4 million today, could be avoided for each of the 227 wide-body aircraft over the life span of that sub-fleet, for a staggering cost avoidance of $454 million, or $1.09 billion today!)

A project was authorized to build the system as described, and I was assigned as the project manager and O.R. analyst to build the scheduling model and algorithm (i.e., a *greedy heuristic* based on *job scheduling on parallel machines with firm due dates* written from scratch in C), along with a software engineer who built a color-coded Gantt chart GUI for the analysts' Macintosh computers that emulated their wall-hanging paper schedule charts of old.

As you might imagine, there was quite a bit of skepticism from the analysts about the ability of a computer to generate higher-yield, more efficient five-year maintenance check, and hangar schedule plan better than they could. Their skepticism turned first to incredulity and then quickly to unmitigated fear and dread when my partner and I delivered an early version of the software within a few months that, with the right input parameters (running on a Mac IIcx desktop computer), could generate a five-year, 600-aircraft fleet maintenance and hangar plan schedule with optimized ~100% check yields in about *18 minutes* (a process that used to take two or three people weeks to generate one feasible plan with 80% check yields)!

At that point, the analysts took me aside and said something to the effect of, "You are going to put the three of us out of work with that computer program of yours!" I not only assured them that was *not* the case but also predicted (and bet them a steak dinner) that they would all get promoted as a result of their ability to use the new system to create more efficient, cost-effective maintenance plans in a far timelier manner than before.

Interactive Optimization

As it turned out, the system we created was very much a case of (AI) *augmentation* rather than *replacement*. The system employed a design framework (conceived at Georgia Tech in the late 1980s, where I earned my M.S. degree) known as *interactive optimization*. The approach combines prescriptive optimization-based techniques, including heuristics when appropriate, and an evaluative simulation-based approach to quickly generate optimized schedules interactively with a human-in-the-loop iteratively providing the necessary inputs and feedback to *guide and push* the algorithm in the right direction toward an optimized solution. Therefore, the human and system work *together*, leveraging their respective strengths to generate better solutions quickly that neither would be able to deliver on their own. Humans can more easily *inspect* a graphical Gantt chart representation of the schedule and see where hangar capacity needs to be added or excess capacity taken away to optimize check yields. A computer can add and subtract, albeit really quickly, and store information, and an algorithm can be programmed to automatically generate maintenance plans and schedules the same way a human would, but far faster.

Suffice it to say the project was a success. My software engineer and I, both working full time, delivered the first production version of the system in about six elapsed months (12 labor-months) and demonstrated how we would achieve the originally targeted benefits over time, i.e., $454 million in maintenance cost avoidance through increased wide-body aircraft check yields, along with multiple additional unforeseen benefits. By optimizing yields and, in effect, pushing aircraft maintenance events out later in time, but still within the Federal Aviation Administration's (FAA) legal limits, the analysts used the model to open up additional hangar space, which allowed:

1. Aircraft maintenance work that had been contracted out to a third party, owing to a perceived lack of in-house hangar capacity, to be brought back in-house and avoid incremental costs.
2. American Airlines should bring in maintenance work from other airlines that didn't have ample hangar capacity (and that work was done at a profit).
3. One narrow-body aircraft to be returned to the fleet and revenue-generating service for a period of one year after an entire maintenance line was deemed superfluous and then shut down.

The analysts not only received promotions but also became a valued, trusted resource to the executives, including the senior vice president, as a result of their ability to "see into the future" with confidence and accuracy and to evaluate all manner of various planning scenarios with the new model/system that they never could have dreamed of doing before.

Keys to Success

What were the key change management factors that made the project a success? There were clear goals, objectives and a well-defined project scope, including a tangible business value target. To start with, I personally spent the first six weeks of the project literally sitting and working side by side with the analysts at the maintenance base in Tulsa, Oklahoma, learning about and understanding the art and science of scheduling aircraft maintenance and hangar facilities, the data and decision-making until I could do the job myself. I listened two-thirds of the time and asked questions the other third.

As a team, we had regular status and update meetings every time my partner and I hit a noteworthy milestone and deliverable (what today we would call Agile-Minimum Viable Product) at each stage of development of the model, algorithm, and schedule GUI. There was ample two-way communication – i.e., we demonstrated what we had done in detail, and the analysts provided constructive feedback and guidance to validate the model's performance and results. I continually reassured the analysts that the system was designed not to operate "completely autonomously" but rather for *them* to operate it and "drive it" iteratively and interactively, much like a driver directs an automobile with inputs from the gear shift, accelerator and brake pedals, and steering wheel to reach their destination.

The changeover from a cumbersome, manual, spreadsheet-based process to a streamlined, automated and interactively optimized process was orchestrated to reduce fear of and instill confidence in the new solution. We endeavored to make the transition to the new system as seamless and stress-free as possible by reusing all of the same data, terminology, scheduling logic, KPIs, report formats and familiar visualization tools in software GUI, such as the

Gantt chart from the historical wall-hung paper maintenance and hangar schedules. That way, the learning curve on the new system was not very steep at all.

The interactive optimization approach, based on the analysts' own step-by-step processes, also made the analysts feel much more comfortable with the solution, rather than being a "black box" that they didn't understand. One of the analysts even referred to the new system as "a big calculator" that could enter the input data and output an optimized five-year maintenance schedule and hangar plan. A great metaphor indeed [3]!

Justification for Change

> One of the things I have learned well from many real-world modeling engagements is that finding the supposedly 'optimal' solution is often not nearly as important as putting the solution values into a form that the client is accustomed to seeing.
>
> *– R.E.D. "Gene" Woolsey, Ph.D., Professor, Colorado School of Mines and Operations Research Academic, Practitioner & Consultant*

The best way to "grease the skids" of change management is to deliver significant, tangible, measurable business value that can be categorically attributed to the new model/solution as demonstrated by before-and-after experiments. We did that in this case, and the "after scenario" delivered far more and better solutions faster than the analysts could have ever imagined. This made for a super easy justification for change. It's rarely that easy, but sometimes it can be. (Promotions, raises and escalation of one's status in the organization goes a long way toward acceptance!)

I went on to use this exact same approach multiple times during my career, including once at another airline to build a new jet fuel supply chain purchasing and inventory management optimization system. As with the maintenance scheduling scenario, the science and technology were sophisticated and substantive, leveraged augmentation versus replacement, and got the job done. That said, it was the "soft skills" that really made the difference. The jet fuel supply chain business folks were quite attached to their 150-tab Excel spreadsheet that they had been using for 30 years, and they did not necessarily want to trade it in for a "new and improved" data-driven, analytically based forecasting, purchasing and inventory optimization model suite. In fact, they initially put up quite a fight. However, when the results of a head-to-head "bake-off" between the spreadsheet and the new models were validated, the supporting case for the new models/system was made: an eight-figure annual cost avoidance opportunity generated by the models in a matter of minutes, versus days and weeks by the status quo process!

The outcome was quite similar with significant business value and economic impact, and satisfied business stakeholders benefited from a continuous close engagement with my team from the start of the project. A smooth, seamless transition and a change management process focused on large doses of communication, mutual understanding and empathy, and iterative testing and validation made all the difference.

Notes and References

1. Machiavelli, Niccolò (1469–1527), 1981, "*The Prince*," New York: Penguin Books.
2. T. Davenport's "The AI Advantage" (2018) does a great job of explaining this concept.
3. Doug Gray, 1992, "Airworthy: Decision support for aircraft overhaul maintenance planning," OR/MS Today, pp. 23–29.

Part 7: Unrealistic Expectations

Under Promise, Over Deliver

– *Tom Peters, TPG Communications (1987)*

One of the critically important lessons that I personally learned the hard way earlier in my career involved setting (un)realistic expectations.

There are two primary domains for expectation setting that plague IT and data science professionals alike:

1. Project-related variables, namely scope, timing, resources, and budget.
2. Business value and economic impact.

Numerous books have been written on Agile Scrum and Kanban estimation, and much of estimation is an art and a science learned over time from lots of practical experience. My only recommendation here is to balance conservatism and stretch goals. It is always better to be a bit early than late, relative to the promised deadline.

In the second domain of business value and economic impact, balancing conservatism and stretch goals is also advisable. My favorite graphic to illustrate this point is shown below.

We establish an *expectation-setting continuum*. See Figure 12.4 for a graphical illustration of the Expectation Setting Continuum explained below.

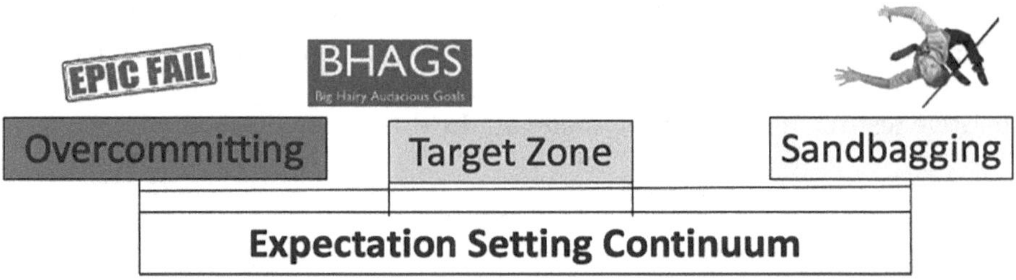

FIGURE 12.4
The art of expectation setting.

At the far right of the spectrum is where we set the bar for benefits too *low*, sailing over the bar too easily and blowing away our target. This approach, known as **"Sandbagging,"** tends to lose a customer's confidence because they perceive the data scientist as not being aggressive enough in targeting potential business value.

At the left end of the spectrum, we set the bar for benefits too *high* and fail to deliver the promised business value. This approach, known as **"Overcommitting"** – or an "epic fail," as millennials might say – can get you into really big trouble with customers (and their bosses/executives) because they were expecting to deliver a monumental economic impact and came up way, way short (e.g., expecting fireworks but got sparklers).

An example of sandbagging versus overcommitting would be achieving a $100 million verifiable cost avoidance but promising $10 million and $1 billion, respectively. With sandbagging, you blow your target away times 10, and with overcommitting, you miss the mark times 10. Both are bad, but overcommitting can be politically irredeemable and career jeopardizing.

In business school, MBAs (of which I am also one) are taught to always analyze (at least) three outcome cases in any analysis or projection modeling scenario:

1. *Best case* (mostly everything goes right).
2. *Worst case* (mostly everything goes wrong).
3. *Expected or average case* (some things go right, some things go wrong, and it balances out).

The process of estimating benefits is similar, and the last case is in the *middle* of the expectation-setting spectrum, in which we try to balance our (and the customer's) optimism and pessimism and use an *expected or average case* to set a target we can hit (or even exceed) without being too far off in either direction. I call this approach the **"target zone."** To continue the example above, we would estimate, say, an $80 million benefit and deliver $100 million. The upper end of the target zone, and just beyond, is sometimes referred to as BHAGs, or **B**ig **H**airy **A**udacious **G**oals (see "Built to Last: Successful

Habits of Visionary Companies" for elaboration on BHAGs). You may have heard BHAGs referred to as "stretch goals."

Finally, to finish the above example, we would set a target zone aim of an $80 million benefit *and* a BHAG of $120 million and deliver a $100 million benefit. Regardless of whether you achieve the BHAG, it is better to aim a bit higher and push for the larger opportunity. Even if you don't achieve the BHAG, you won't fall as far short by overcommitting as wildly as with the $1 billion benefit.

Clearly, the examples above are contrived because we have the benefit of hindsight, or perfect information, and we achieved a healthy business value outcome of $100 million. That does not always happen in reality. Sometimes the benefit is $0 or something close. Sometimes we get lucky and hit the jackpot.

Setting Business Value Benefits

I'll provide some key learnings on setting business value benefits and economic impact targets *before* a project commences.

1. The size of the business, in scope and scale, measured in terms of sales, revenue, costs, assets, labor force and profits, greatly matters in how much business value can realistically be achieved in general and in any one project within the business.

   Some of the greatest achievements in the history of data science (operations research, analytics) at Fortune 50 companies were nine-figure annualized business value improvements:

   - American Airlines' yield management system (DINAMO) was validated to have generated $1.4 billion in incremental revenue over a three-year period (and was expected to deliver ~$500 million in incremental revenue annually in the future) with a similar fleet and airline schedule (see "Yield Management at American Airlines").
   - UPS' On-Road Integrated Optimization and Navigation System (ORION) avoids $300 million to $400 million annually in fuel and driver costs (see "Meet ORION, Software That Will Save UPS Millions By Improving Drivers' Routes").

     Although there have been many examples of even larger annual business value benefits of data science and similar fields, the 9- to 10-figure dollar range provides a reasonable upper bound on the largest possible practical expectations.

2. The best place to start is with the firm's financial statements to understand financial performance to estimate benefits, and then examine a given department's contributions to the firm's financial results (Marketing, Manufacturing, etc.).

Key areas of the business for economic opportunity include:

- *Labor*, e.g., through AI-based robotics in warehouses, factories and, in the future, driverless vehicles, including cars and large trucks, as well as optimization-based labor planning systems.
- *Inventory*, e.g., better matching of product demand to supply to balance shortage and holding costs.
- *Asset* (including facility) allocation and utilization, e.g., aircraft (hangars), railroad engines and rolling stock, and tractor trailers.
- *Manufacturing*, e.g., product mix, process controls, statistical quality control.
- *Pricing*, e.g., Walmart was a 2020 INFORMS Franz Edelman Award finalist with the predictive (demand forecasting)-prescriptive markdowns optimization solution that balanced discounting goods too much or too little, too early or too late, to maximize sales revenue.
- *Yield or revenue management*, e.g., originating in airlines and now utilized in hotels, cruise lines, rental car companies and even self-storage facilities.

3. When estimating benefits, first calculate the maximum potential benefit (at 100% realization), and then perform a rigorous analysis based on the firm's actual economic data to evaluate how much business value and economic impact is realistically possible to achieve with data science.

   The latter figure may very well only be 10%–25%, and you may decide to set a target zone goal at 5%–10% of the maximum to avoid sandbagging or overcommitting. Rarely, if ever, will you achieve the maximum benefit, but multiplying 10% times a very, very big cost or revenue dollar figure can still be a significant number in itself.

4. Look for the largest potential business value opportunities in your company where there is significant room for improvement in economic efficiency to see how data science can provide the greatest leverage. Look for complex, large economic impact problems that are currently being solved essentially manually in Excel with rules of thumb or simple heuristics that do not capture the fullness of the problem or solution opportunity.

   In the airline industry (American, Delta, et al.), the largest opportunities were found in seat inventory pricing and yield management, which directly impacted revenue, followed closely by network planning/flight scheduling and flight/cabin crew scheduling and fuel inventory management, which are an airline's two largest cost categories, followed by spare parts inventory and aircraft maintenance.

In the package delivery industry (UPS), the largest opportunities were found in optimizing the operations of their fleet of 55,000 delivery trucks and drivers (not making left-hand turns because you burn more fuel waiting to turn!).

Setting realistic business value and economic impact targets and expectations will depend on how well you understand the economics, operations and financials of your company and then how rigorously you analyze the impact that data science can potentially have by utilizing the framework of target zone and BHAGs from the expectation-setting continuum.

Part 8: Project Management

Nine women cannot have a baby in one month.

Brooks' Law: Adding resources to a late software project only makes it later.

The Standish Group has published their "CHAOS Report" for nearly 40 years, chronicling the failure of most IT projects to achieve scope, timing, budget *and* quality goals (all four). As of their 2020 "CHAOS Report: Beyond Infinity," only 19% *of all IT projects* achieve these four lofty goals, which has not gotten much better in the past 40 years and, frankly, may even be worse. Many projects will, of course, achieve a combination of some subset of these goals, e.g., timing and quality but not scope and budget.

Why is this statistic even relevant in the context of data science project failures? Beyond the data analysis and modeling phases, a.k.a. the "fun part," successful data science modeling projects ultimately evolve to become software and systems engineering projects. We recall Tom Davenport's "Begin with the end in mind" goal: "Models make the enterprise smarter; models embedded in systems and business processes make the enterprise more economically efficient."

(I will delve more into the complexities and complications of transforming a *model* into a *turnkey business system* for planning or real-time decision-making in a later installment.)

Building a turnkey production system-based version of a model that supports an enterprise business process of more than modest importance and criticality will entail building API interfaces to multiple data sources, architecting a microservice application around the model, built-in model drift detection, refitting the model, model assumption revalidation, error handling, fault tolerance, high-availability and high-reliability robustness, and failover capabilities to get to 99+% uptime (mission-critical systems require the elusive "5–9s" or 99.999% uptime). Having been through this process many, many times myself, strong project management (PM) practices, processes, skills, discipline and judgment are critical to successfully achieving scope, timing, budget and quality goals, or at least not the null set.

The "Perfect" Project Management

PM for modeling initiatives is quite a bit more relaxed in that it is understood to be a "research project" – an application of the scientific method, a voyage of discovery, almost as prone to "puffs of smoke over the lab bench" as Edison inventing the lightbulb. The four dimensions of the inviolate PM "box" (formerly a triangle until the addition of "quality") are no less important to set realistic estimates and expectations for stakeholders as to how long a project is going to take. See Figure 12.5 for a graphical illustration of the four-dimensional PM "box", i.e., scope, timing, budget/resources, quality.

Agile/MVP (Minimum Viable Product), and its precursors extreme programming and rapid prototyping, has provided a marked improvement over Waterfall *(Egad!)* in creating a more flexible, realistic framework for software and system engineering projects. *(According to The Standish Group, software projects using Agile methods are three times more likely to succeed than those using the Waterfall method, and Waterfall (software) projects are two times more likely to fail.)*

Although Scrum is more popular for strictly software projects, I tend to prefer and recommend Kanban for modeling projects because modeling is more reactive in nature and data scientists are continually discovering new variables, constraints and data sources with which to integrate, etc. The notion of MVP, or Minimum Viable *Model*, is most appealing to emphasize the importance of stakeholder confidence to get a basic model up and running and to functionally provide some insightful output ASAP.

Once your model is deemed successful and too important to live without reliably, and a budget is approved to convert it into a production system

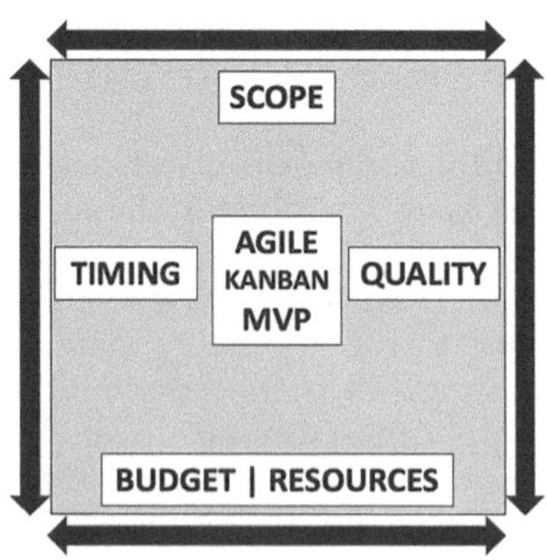

FIGURE 12.5
The four dimensions of project management.

(mission-critical or not, planning or real-time), you are going to become part of a much larger team consisting of data engineers, software engineers, cloud data center system engineers, test and QA engineers, project managers, business analysts, etc. Most likely, you will no longer be "in control" of the project.

Teams will typically employ Agile Scrum for systems development and SAFe (Scaled Agile Framework) for enterprise-scale systems development. I have used Agile Scrum, Kanban and SAFe, and I highly recommend *training* in all of them to reduce project management risk.

Project management is at least *part* science (refer to https://www.pmi.org/certifications/project-management-pmp), and there are many tools and techniques available for managing software projects. That said, project management is also an art form that is based on experience, instinct, judgment and most importantly, knowledge about your people, processes, technology and business problem domain. The fewer unknowns and "new stuff" on a project, the higher probability for success. (But refer back to the 2020 CHAOS Report for a reality check and the reasons IT projects fail.)

The Usual Suspects in PM

There is a set of "usual suspects" as to why IT and data science projects fail:

- Underestimating scope.
- Overcommitting on scope.
- Underestimating complexity (e.g., technological, architectural, change management, and system integration).
- Overestimating team capacity and capabilities, especially new staff or newly formed teams (there is always a ramp-up, learning curve period).
- Overprioritizing too many features (i.e., every feature cannot be Priority Level 1 for Release 1).
- Unrealistic timeline to address scope and no slack in the timeline for contingencies and unforeseen circumstances.
- Insufficient quantity of (adequately skilled) resources to address scope.
- Project manager's ability to succeed through adversity and make decisions to course-correct when things go awry (and they ultimately will).

Like most endeavors, e.g., learning to play an instrument or a sport, PM skills are developed through the experience of *doing* project management, including making mistakes and learning from them, not reading about it in a book or taking a course. Studying project management may be necessary but is by no means *sufficient* for developing and honing your skills and becoming a good project manager. There is no substitute for PM experience or learning from other, more experienced project managers. In my experience, it takes years of hands-on intensive experience managing projects of larger and larger scope and greater and greater complexity to develop expertise.

The best advice I can provide is to always err on the side of caution when making project commitments on scope, timeline, resources, budget, quality, and complexity, and when in doubt, seek the advice of more experienced project managers. Always be transparent with your team members, stakeholders and constituents, and report bad news and offer solutions as soon as you discover an adverse situation. *(Bad news does not get better with age!)*

Below is a generalized, representative, phased data science project lifecycle that I have developed based on decades of real-world implementation experience by professional data scientists that you can utilize as a template to help guide your own project planning process and expectation setting. *(Your mileage may vary.)*

Part 9: Excessive Focus on the Model, Technique, or Technology

> A model is a means to an end, not an end itself.
>
> I honestly do not understand all of the math, but I am convinced of the strategic competitive advantage, and significant, tangible, economic value that is created with Yield Management.
>
> *– Robert L. Crandall, Chairman, President, CEO of American Airlines circa 1989 (Wharton MBA)*

I, for one, have always been enamored with the power and beauty of mathematics. The notation is a language unto itself. Known as the "Queen of the Sciences," mathematics provides the tools to enable other sciences, such as physics (which provides the foundation for all of the engineering fields) and economics.

In capitalism, businesses are in business to make money and return that money to shareholders while benefiting society along the way.

At the intersection of mathematics and business, fields like operations research, management science, statistics, and now, analytics and data science are intended to contribute to the betterment of the corporation's economic and financial performance. *The mathematics and models are a means to an end, not an end themselves.*

It is not uncommon, especially among recent graduates, to become excessively focused and a bit too enamored with the model and mathematics, the algorithms and technology.

The Pareto principle (80/20 rule) can be of interest and application here, i.e., getting 80% of the benefit for 20% of the effort (or cost). *Perfection is the enemy of done!* In business, most of the time, there is no need or willingness on the part of management to expend that 80% of the effort to gain the last 20% of the business value. The business needs an answer … and value delivered … *now*. It doesn't need to be perfect. It just needs to work and deliver against the economic impact objectives.

The Agile principle of Minimum Viable Product (or Model) is directionally correct and applicable as well. *Get to a version that builds, works and generates value ASAP.* See Figure 12.6 for a representative illustration of a Data Science Project Lifecycle including typically encountered estimates of Project Phase (I-V) activities and durations, and the total elapsed time to begin generating business value on enterprise-scale ADSAI projects.

In most businesses with which I have worked, the goal was for *minimal elapsed time possible to value realization.* In fact, in status reports that go to senior management, project update entries *must have a business value attached*, or they are omitted forthwith. No technical jargon or detail is even allowed. It is implicit and assumed that the correct model form was utilized, tested and validated, as was the business value.

Excessive tweaking, refinements and feature additions or modifications, for little or no measurably incremental gain, are a waste of the company's time and resources.

I had a team of EMBA students whose final project in a Business Analytics course was focused on improving the accuracy of (binary classification) models to predict mortgage loan defaults operating on a large volume of historical loan performance outcome data (a technique that would have come in handy circa 2008–2009). The students filled 20 PowerPoint slides with mind-bending mathematical models, arcane terminology and symbols, and spent 19 of their 20 allotted presentation minutes talking about all of the different mathematical and statistical models that they had built – Fast Fourier Transforms, Bayesian inference, neural networks, etc. In minute 20, I finally raised my hand and asked, *"What was the* ***business outcome result*** *you achieved?"* They responded, *"Oh, wow, we increased the accuracy of mortgage loan default prediction to over 93%! On their typical loan portfolio, the new and improved model was going to avoid tens of millions of dollars in bad loan default write-offs annually!"* To which I responded, *"In the future, when presenting, especially to executives, please start with that information."* In journalism, this practice of omitting the most important pieces of information is called "burying the lede."

The moral of that story should also be applied when presenting to executives inside your company. No one, except perhaps other mathematicians at a conference, cares about all of the technical details. Save that for the appendix. Instead, tell a story of what life was like *before* and *after* the model was implemented. Focus on the improved business solution and the incremental business value and economic impact that was achieved in terms of cost, revenue, asset utilization and customer satisfaction – things they understand in terms

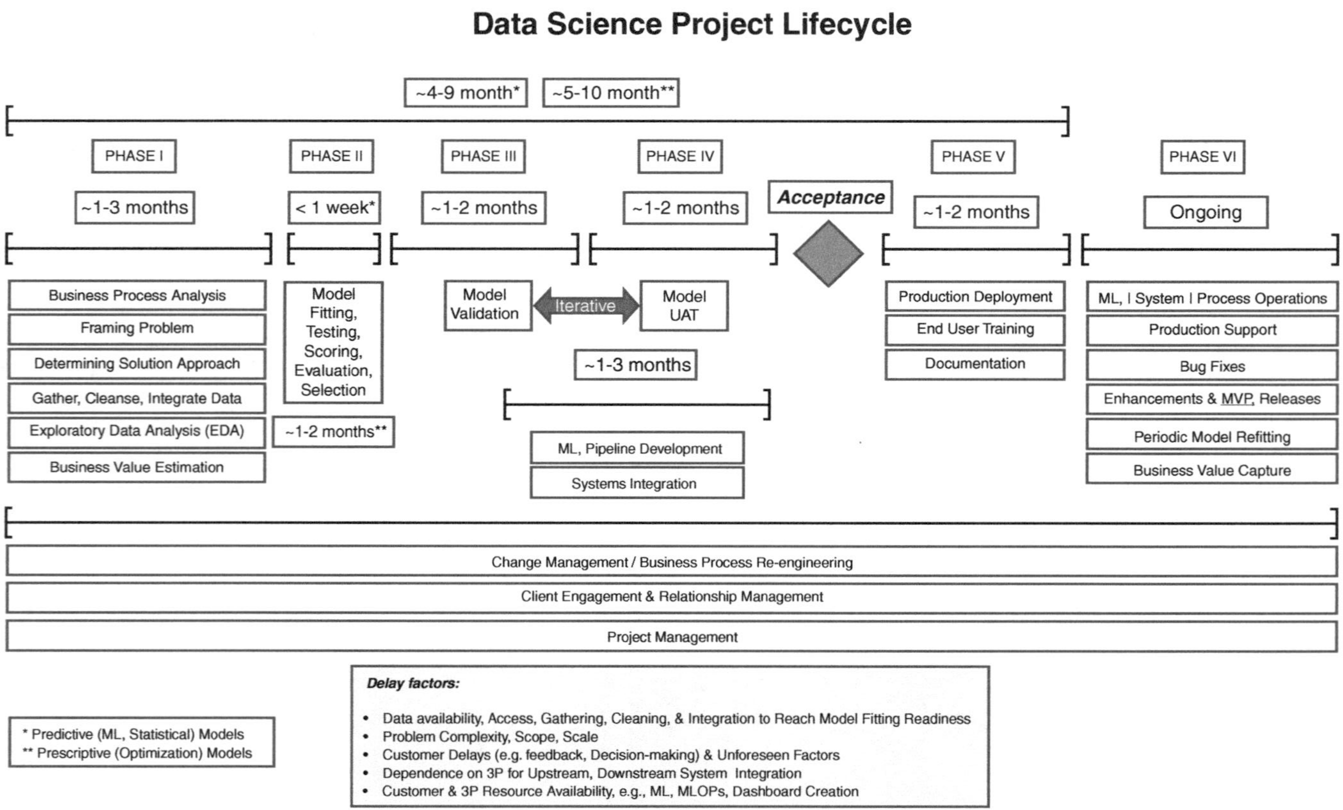

FIGURE 12.6
The lifecycle of a data science project.

they use every day. Explain how much time and effort will be saved with a streamlined, automated process. Refer to the controlled experiments that were run to prove out the model's value, and the testing and validation, with the business domain and finance folks. They'll want to know that you can "show your work," but they don't need to see or hear about all of the math and code.

Part 10: Getting from Sandbox Model to Production System

> Building models is easy ... you can do that in a day, especially now with AutoML, or in weeks with optimization suites ... building a model into an enterprise production-grade system is difficult ... that can take years ... and cost millions of dollars, especially if it is mission-critical or utilized for real-time decision-making.

How long does it take to transform a model from the sandbox to become a production system? The answer, as usual in business (and consulting), is, "It depends (on many factors)." It can take months or sometimes years for large-scale, sophisticated systems to solve the most complex types of problems that must be available, be reliable, and access large amounts of data across the enterprise.

Some of the most significant factors to consider and gauge when determining complexity include:

- *Dynamism* – Is the input data largely static, such as planning information, or continuously changing in real-time? (System complexity increases as you move toward real-time.)
- *Integration* – Is the system relatively standalone or heavily reliant on integration with numerous other data sources and enterprise systems? (System complexity increases with the number of integration points.)
- *Mission Criticality* – If the system can fail and no one is more than moderately inconvenienced, that's easier; if the entire company grinds to a halt, then multiple layers of error handling, fault tolerance, and failover capabilities and infrastructure must be built in to increase availability and reliability (99.999%), which becomes far, far more challenging.
- *Problem and Model Complexity* – How difficult is the underlying problem, and how sophisticated, mathematically and computationally, is the model generating the solution?

The first project I worked on from the "ground floor" (a bare C++ compiler screen; see Part 6) was to build American Airlines' aircraft maintenance check/hangar planning and scheduling system. It took me and another software engineer working in parallel about six months to deliver a fully system-tested, working minimum viable product (MVP) Version 1. It took another three months to install and configure the software for the local environment, train the users and work out some additional bugs and kinks discovered during a rigorous user acceptance test on all fleet scenarios (e.g., it turned out 1 of the 3 was color blind and needed a different graphical user interface to differentiate different-colored screen icons).

The system was intended for long-range (five-year) planning and scheduling purposes, not real-time decision-making on operational considerations. The main input data file from which the system was driven that reported updated accumulated flight hours by aircraft tail number was downloaded from a mainframe and uploaded into the application once per week. All of the other input files (about a dozen) were manually edited, kept up to date, and modified to run a variety of scenarios. Using these input files, the system generated a schedule in a matter of minutes that was displayed on a computer in the form of a scrollable five-year Gantt chart.

Thus, the level of dynamism was low, the level of integration was low, the system was not mission-critical (worst-case requiring a PC reboot), and the problem and model complexity were low (a greedy heuristic algorithm solving a well-understood job scheduling problem on parallel machines with firm due dates for 600 aircraft divided into subfleets). Net-net, this was a relatively easy problem to solve and system to build and deploy in less than a year in an enterprise departmental context.

At the other end of the spectrum, consider some of the systems I've mentioned in previous installments, including:

- United Parcel Service (UPS) **ORION** (Delivery Truck Routing).
- American Airlines (AA) **DINAMO** (Yield, Pricing & Seat Inventory Management).

Both of these systems have a high level of dynamism, a high level of integration, a high degree of mission criticality and a high degree of problem complexity and solution sophistication.

UPS' ORION is solving 55,000 traveling salesman problems (which is NP-complete) (one for each delivery truck) based on package delivery information scheduled for the next day. The system took more than 10 years to build and deploy and cost $250 million, but saves $300 million to $400 million in costs annually.

AA's DINAMO was performing 22,000 passenger flight demand forecasts nightly and dynamically optimizing seat inventory pricing using a mixed-integer linear programming model for thousands of flights and 600 aircraft per day (circa 1991). The system took several years and cost hundreds of millions of dollars to build and deploy (billions if you count Sabre, the underlying computerized reservation system) and generates an additional $400 million to $500 million in revenue annually.

With colleagues Dr. Phil Beck and Dr. Mark Song, and domain expert Supervisor of Dispatch Charles Cunningham, I helped lead the successful development and implementation of a real-time airline disruption recovery optimization system at Southwest Airlines that had a high level of dynamism, a high level of integration, a high degree of mission criticality, and a high degree of problem complexity and solution sophistication.

Code-named "The Baker" (posthumously named for Supervisor of Dispatch Mike Baker, the domain expert who originally conceived of the concept), the system could take a real-time, data-based "snapshot" of the airline (i.e., aircraft, flights, passengers, maintenance events, airport weather and operating conditions, such as low ceilings, curfews or ground stops, but not crew members) before, during or after a major disruption, like a snowstorm, hurricane, FAA Air Traffic Control (ATC) ground stop or other, and use a network optimization algorithm with side constraints to recommend flight delay and cancellation decision alternatives with the intent to minimize passenger disruptions. Solutions could be generated in 5–30 minutes for isolated or network-wide scenarios, respectively (such complex manual decisions required several hours, at best, by which time, conditions would have changed, further complicating decision-making).

Recognized at the time by senior airline operations executives as the "single most influential operations-oriented system ever delivered by Technology at Southwest," the system took about eight years to complete (from initial concept in 2008 to initial delivery in 2015, mostly because of the quantity and real-time nature of the input data required via integration with Southwest's primary airline operations information system which was also under development concurrently with The Baker). The Baker cost millions of dollars (mid to high seven figures) to develop and deploy, dramatically improved the airline's On-Time Performance (OTP) in major winter storms by 2x and cancellations by 2x and increased overall airline OTP by 2.11 full percentage points. The number of passengers delayed by two or more hours decreased by 95% when The Baker was fully implemented in 2016.

This project succeeded when many, many prior attempts by other airlines had failed to make even a dent in this incredibly complex, real-time airline operations decision-making problem. That success is a testament to the hard work, partnership, commitment, perseverance, business domain knowledge and skill, and superior technical excellence of the team members that The Baker a reality!

Failure Begets Success

In 1990 at AA, O.R. staff had great success building and successfully deploying limited day-of operations control analytics models into production, e.g.,

Arrival Slot Allocation System (ASAS) (a 1990 Edelman Award finalist), which optimized assignments of arriving AA aircraft to airport-specific FAA ATC arrival positions or "time slots" according to the flights' actual versus scheduled estimated times of arrival (ETAs). However, when my colleague and I attempted to expand the scope of these models to network-wide flight delay and cancellation and aircraft swap decision-making, we did not have all of the real-time or updated data or computing power required to solve such large-scale problems. We wrote down a very elegant mathematical model formulation, but when we went to implement it, we failed due to a lack of streaming, real-time data and computing power – net-net, the project never got beyond the prototype stage. The head of AA Systems Operations Control at the time wasn't enthralled with the idea of turning over that level of real-time airline decision-making to a "computer" – the change management hurdle to continue the project was just way too high at the time.

These examples of successful and failed projects help to illustrate that there is a broad spectrum when transforming from a sandbox model to a production system. Factors including dynamism, integration, mission criticality, and problem and model complexity play a huge role in determining the time, effort, resources and investment required to deliver an enterprise system.

That said, smaller-scale, relatively simple systems implemented with relatively little investment can still deliver tremendous business value compared with incumbent solutions, if the economic potential is there. However, large (to extremely large) relative investments are required to reap the enormous rewards delivered by robust, larger-scale, more sophisticated systems.

Therefore, truly understanding scope and complexity prior to delivering estimates or embarking on such a journey is critical to success.

It is advisable to keep the following in mind:

- Building production systems around and underneath models is 10–100 times more complex and resource-intensive than building the model itself.
- The best bet is to build the model as a microservice that attaches via contract-programmed APIs rather than a standalone system; chances are, in most large enterprises, a model will be a part of an ecosystem of many other systems rather than a standalone entity.
- The model will need support from the Data Organization, in particular Data Engineering and Governance, to provide the data pipelines needed for access to timely, high-integrity data and their sources.
- The model will need support from the Technology Organization, in particular – software engineering – to facilitate the interfaces to other enterprise systems; cloud services to provide the compute, storage and network services needed; and test/QA to verify, validate and certify the model's application.

If your company has a Change Management team, I recommend enlisting their help with orchestrating the transition of all involved business constituent staff and management stakeholders from "the old way of doing things" to the new world order that your model is creating. Change management has become its own discipline and can help with major changes in policies, processes, and procedures driven by your model.

Conclusion

The goal of data science, and related fields like analytics, is to help solve complex strategic, tactical, and operational problems, support and better enable data-based, model-driven decision-making, and answer key business questions in such a manner that business value is created and economic impact is maximized. Tom Davenport set the bar necessarily high when he said, "Models make the enterprise smarter; models embedded in systems and business processes make the enterprise more economically efficient." For data scientists, this is our desired end state, and the end in mind with which we begin our endeavors.

Like any scientific endeavor, experimentation and data collection and analysis are a part of the process, combined with the use of advanced mathematics and sophisticated software and computer technology. Notwithstanding all of this science and technology, the practice of data science takes place in the private sector (e.g., business, industry, research) and public sector (e.g., government, military, law enforcement), all of which are inhabited and operated by human beings.

Human involvement in data science is substantial in every step of the process and materially significant, requiring the development and application of many "soft skills" that necessarily facilitate successful execution and completion of data science projects.

Of all the soft skills required, I believe that *communication* is by far the most critical and foundational to successfully execute data science projects. Communication in all forms, specifically:

- Listening, to understand
- Being heard, to be understood
- Speaking and writing concisely, impactfully and with clarity for the target audience
- Gaining a deep level of mutual understanding in all aspects of a project

Communication is critical for all parties involved, especially the data scientist and business manager ("customer"), as well as others including data engineers, software engineers, business analysts, data governance analysts and potentially others. Many different individuals, with their respective unique skill sets and perspectives, are often required to complete a data science project. Clear, concise communication and mutual understanding and agreement among all stakeholders on all critical facets of the project are essential for success. If there are n people on a project, the number of possible communication links between them are $n(n-1)$, or on the order of n^2 ($O(n^2)$). It's no wonder Brooks's Law states, *"Adding resources to a project that is already late will only make it later"* ... communication alone is half the battle.

Project Challenges

Understanding the business problem that you are trying to solve is often a point in which data science projects go awry. Sometimes the business folks themselves are not completely clear on what the real problem is. Therefore, we should not be surprised that the data scientist may need to do quite a bit of investigating, along with the business folks, to determine the problem to be solved. Sometimes, the business folks *do* understand the problem, but there is a breakdown in communication, such as lack of a clear explanation from the business or failure to adequately listen and ask clarifying questions on the part of the data scientist, that inhibits mutual understanding of the problem. Getting to a clearly stated and mutually understood problem definition, as well as associated business process flows, data flows and decision-making processes and criteria, is foundational to initiating and successfully completing a data science project.

The challenges associated with data are many and will continue to hinder data science projects. Historically, challenges include not having enough data or not having it in one place for analysis, and going forward, having too much in too many forms and in too many locations. Great strides are being made in the fields of data engineering and data governance and the development of technology platforms that support these endeavors. The volume and dynamism of data generated by myriad enterprise systems, e-commerce and social media platforms, IoT devices, etc., will continue to generate more data than most enterprises can realistically, let alone easily, manage. The key for successful data science projects is to focus on the data that you must have for your project to get to MVP/M (Minimum Viable Product or Model). You can always add in relevant data when it becomes available down the road.

Misapplying a model often occurs when faulty or improper assumptions are made about the applicability of a particular model form or its usage to solve the problem at hand. Experimental design is a critically important skill that is often lost on citizen data scientists, and some professional data scientists, and is usually attributable to a lack of training and education in the subject. Although techniques can be quantitatively applied, there is also an artfulness to a well-designed, statistically valid experiment. Predictive model bias and overfitting are also common errors that result in invalidated results but can be avoided with properly applied techniques, e.g., k-fold cross-validation. When in doubt, consult with a professor or more experienced colleague, and check your textbook and online references to ensure that the model you are employing is valid, and the experiment you are running is suitable for the problem at hand. (Google is your friend because it is highly unlikely that you are the first person to encounter a given type of problem, or one that is similar. A thorough literature search is encouraged to model development.)

It should go without saying that business folks and data scientists should focus on problems and data science projects that represent an agreed upon (high) business priority, i.e., ones that will realistically generate significant business value and economic impact; however, you measure it. Unfortunately, that is not always the case. Sometimes business folks do not have a clear set of prioritized projects ranked on true business values. Even if they do, sometimes data scientists, and, yes, even business folks, get distracted by other initiatives that consume time and resources. The *Key Business Question Grid* and the *Project Valuation Ranking* tool in Part 4 can assist in focusing on the highest priority, highest potential problems and projects.

Managing Change

Data science projects induce inordinately large amounts of change. Data science fundamentally and even radically changes the way that problems are solved, questions are answered, and decisions are made. In general, the transition to becoming a fact-based, data-driven enterprise is transformational and fraught with many dimensions of change, including moving away from gut instinct and Excel-based heuristics and rules of thumb to more rigorously rational model-based approaches to complex problem-solving and decision-making. Data scientists may lead the way, but everyone must go on the journey together. Data scientists may inform and teach others how these advanced techniques and technologies function, but everyone, from analysts to managers to executives, must "buy in" to be successful both on individual projects and the overall transformation driven by data science methods and stakeholders.

Communication plays a critical role in managing change and "winning hearts and minds." Storytelling using *before and after* comparisons including lots of *data visualization* to highlight the business impact generated by data science model-based solutions is crucial to demonstrating and proving to management the efficacy of these scientific approaches. (Most people love stories, with lots of pictures to help understand complex topics.) Opting in to *augmentation*-based AI methods and *iterative interactive optimization* approaches can ease the transition from the exclusively human- and Excel-centered approaches to problem-solving and decision-making to the compelling alternative founded on the greater analytical rigor offered by data science. Everyone in the stakeholder/constituent group must be convinced beyond a reasonable doubt that all of the (sometimes painful) changes driven by data science are made worthwhile by the business value and economic impact achieved. ("The juice is worth the squeeze!")

No one likes to be disappointed, or the letdown that follows. Everyone wants a big win on their data science project to help the company and stakeholders, and further their own career progression. All the more reason to set *realistic expectations* on all of the relevant KPIs for the project, i.e., scope, timing, budget and business value targets. Leaning toward conservatism is the best approach and usually wins the day. *Stretch goals* are fine, as are BHAGs (Big Hairy Audacious Goals). Epic fails, caused by over-promising and under-delivering, could ruin a career (or could even get someone fired). Sandbagging draws skepticism and leaves the constituents with a lack of trust.

Project management (PM) has evolved to be considered both a science and an art. Techniques such as PERT/CPM and Agile burn-down charts attempt to quantify and measure how a project is progressing and how well a team is performing. These tools are invaluable to a project manager, but they primarily *inform*. There is a disproportionate amount of *judgment* that must be applied to managing the scientific modeling aspect of a data science project, as well as the systems development activity. Measuring and gauging *complexity* of a task that will impact resource consumption and timeline is a skill that comes from the experience of working on numerous projects with wide-ranging high and low levels of difficulty and learning how to approach and solve for them. Measuring and gauging a *team's output and productivity level* as it rises and wanes over the course of a long project is a skill developed through both observation, informed by data, and active interaction with team members as they climb learning curves and struggle to overcome a series of challenges with data, changes in scope, infrastructure issues and more. Project managers who know when to press or ease off and by how much, and when to challenge or relent, are a rare and skilled breed of professional that evolve from experience over time, not from PM certification courses alone.

As painful as it is to admit after spending years and years studying and learning all of that mathematics of far more than modest rigor, and learning to write and test code, no one, other than you, your data science colleagues, and your professors, care about the model, techniques or technology. People in business, and the higher up the leadership chain they are exponentially, care more about the business value and top and bottom economic impact of your data science than the math or code. They trust that you "did the math," but they don't want to hear about it. Sorry.

My advice is to not let yourself get "wrapped around the axle" with a lot of nuanced, overly sophisticated mathematics for its own sake on corporate data science projects. Please, do yourself and your constituents and stakeholders, a *BIG* favor, and *save the math and code* for the Appendix of your presentation, for industry conferences and symposia, refereed academic journal publications, and your Data Science Center of Excellence and community of practice meetings. Always remember the Pareto principle (deliver 80% of the value for 20% of the effort), Minimum Viable Product/Model (MVP/M), and that *perfection is the enemy of done.* The model and code need to be verified and validated, but not perfect.

The final and highest hurdle to achieving data science project success is getting your model from the sandbox (of your desktop or cloud-based work area) to a full-fledged production system (e.g., microservice or standalone) embedded in a high-value business process. Availability, reliability, and repeatability are necessary for your model/system to achieve the "flywheel" of continuously ongoing business value creation without regular human intervention. This process/journey requires a *team* to realize the endgame – business people (i.e., executives for funding and political "air cover," line managers to drive change, and individual contributors to help design, develop, test, validate and implement the solution), technology people (i.e., software, cloud, security, etc.), data people, test/QA people. It may take months, years or even a decade, and may cost hundreds of thousands, millions or hundreds of millions of dollars to deploy and implement completely, depending on the scope and complexity of the problem and the level of sophistication and operational criticality of the model/system solution. (It is advisable to make sure that the benefits delivered by your solution are proportional to the costs to build and implement the same, by whatever measures and metrics the finance department/board of directors utilizes, e.g., NPV, IRR, and ROI.)

As with any human endeavor involving teams of people, whether it is a co-ed softball team, delivering a data science project, or an expedition climbing Mt. Everest, *empathy* is the most important quality to embody regardless of how incredibly difficult things get along the journey. And trust me, as worthwhile as data science projects are in every respect, things will get difficult at many, many points along the way, and you *cannot* afford to alienate any constituents, partners, teammates, or stakeholders. People never forget heroes, and they never forget jerks. You may (barely) get through one project, but you will never get through another one by treating anyone who matters with anything less than The Golden Rule.

A few tenets of advice that go along with empathy when times get tough include:

- Assume positive intent.
- Give people the benefit of the doubt.
- Put yourself in the other person's position.
- Trust, but verify, until people prove themselves unworthy of your trust.
- Delete the angry email before you hit SEND (better yet, don't even write the email, go have a diplomatic conversation).
- Think, then breathe deeply, before you speak.
- Work the problem, don't blame the person who created or uncovered it.
- Read "Emotional Intelligence" by Daniel Goleman for great advice on your EQ relative to your IQ, and the attributes of great senior executive leaders.

To engineer is human. Failure is feedback. Through failure, we learn to succeed.

I sincerely hope that this 10-part series on why data science projects fail will help you and your company to be more successful in all of your future data science endeavors.

13

Surrounded by Success

Introduction

In the late 1990s, the combined field of operations research/management science (OR/MS) was having a "crisis of confidence" in its collective ability to deliver meaningful, real-world business value results and economic impacts. (Since then, OR/MS has undergone a branding and marketing transformation to be closely associated with the field of *analytics* and is application- and mission-wise adjacent to and closely aligned with data science and AI – including machine learning, of course – which evolved from statistics and computer science, respectively.) At the time, the OR/MS discipline was still evolving from being a largely *theoretical* discipline dominated by academics with heavy emphasis on mathematics of more than modest rigor toward a more *applied* discipline embedded in companies in complex industries, such as transportation (e.g., airlines, railroads, package delivery), energy, and financial services.

Peter Horner, then-editor of *OR/MS Today*, the flagship magazine of INFORMS (The Institute For Operations Research and Management Science), asked me to write a commentary from the viewpoint of an OR/MS practitioner who had worked in and outside the field to provide a perspective on where to find opportunities for success. Being the eternal pragmatic optimist that I am, I obliged the request because I had witnessed first-hand the extraordinarily widespread success of OR/MS at American Airlines and Sabre during the initial 10 years of my professional career. I provided examples of successes I experienced working in the OR/MS field at American Airlines, as well as other successful, albeit smaller-scale, analytical projects my team and I implemented after I had transitioned to leading Sabre's internet e-commerce travel reservations software technology platform team (for Travelocity, American Airlines, and multiple other customers).

Given the recent explosion and ubiquitous pervasiveness of applications of analytics, data science, and, most recently, AI (think ChatGPT LLMs and robots), such an article seems hardly necessary today. Thanks to an abundance of rich data from e-commerce, IoT, and enterprise systems, and practically unlimited computing power (think cloud and GPUs), these fields and

DOI: 10.1201/9781003588344-14

their application permeate every department within many, many companies, from marketing, finance, merchandising, supply chain, operations, manufacturing, to even, yes, HR!

The key insights in the article that foresaw the success of OR/MS, now analytics, data science, and AI, and are still very relevant today, including:

- Framing and visualizing business problems through an analytical "lens" to identify opportunities for value creation through modeling and model-based solutions deployment
- Focusing on business value and economic impact, i.e., reducing costs, increasing revenue, enhancing customer service, improving efficiency, in each and every analytical project
- Engaging with the business to deploy solutions embedding the same within production systems and business processes
- Recognizing there is no "silver bullet" for success and no substitute for getting your hands dirty, a lot of hard work, combined with a diligent customer focus, excellent communication skills, and most importantly a measurable, demonstrable value-add
- Validating the business value captured and communicating it with business stakeholders
- The best measure of success for OR/MS professionals is getting asked back for another assignment, and then a continuous stream of projects, as evidence of the value created and captured for customers

When projects are well-executed and deliver value, you will literally find yourself surrounded by success.

THE PROFESSION'S PARADOX: SURROUNDED BY SUCCESS

I often hear and read about people bemoaning the death of the operations research and management science profession. To these people, OR/MS seems to suffer the indifference of a world too busy to care about the value that mathematical models can bring to better understand what makes businesses and organizations operate more effectively and more efficiently. Many OR/MS professionals seem to be at a loss as to how to create and declare success.

I think that depends on how you define success. Is success deriving some new elegant algorithm? Is success having the CEO of your company stand up and say your OR/MS work is truly amazing? Is success winning the Edelman Award, the INFORMS Prize or the Lanchester Prize? Although desirable ends, these are pretty narrow measures of success by which most of us are doomed to fail. However, if you define success as using OR/MS effectively in your daily work and contributing to some value-added positive end – regardless of the endeavor – I believe that we can have many more OR/MS successes, albeit on a smaller-scale, in addition to the large-scale ones. There are more of those seemingly "little successes" that ultimately have a larger positive impact than most people realize.

Successful OR/MS applications are all around us, if you know where to look. Plenty of good OR/MS practice is hidden by job titles. Lots of people do OR/MS, but don't carry the title OR/MS analyst. I know professional people, e.g., market researchers, investment analysts and project managers, who effectively apply OR/MS methods in their daily work to achieve significant, positive results despite the fact that their primary educational training wasn't in OR/MS. Product managers, who must balance the cost to create a new product with market share objectives to determine a competitive price, apply fundamental OR/MS principles. I have one friend who is a manager of information technology for a local police department who writes computer programs that do everything from generate and analyze crime statistics to optimize patrol routes and officer allocations. OR/MS is where you choose to find it.

Alan Greenspan, chairman of the Federal Reserve Board, recently spoke to Congress at length on C-SPAN about the econometric models he uses to reveal the mysteries of the economy, specifically the Consumer Price Index (CPI), which attempts to measure inflation by analyzing changes in the prices of basic consumer goods. However, he also talked about his "personal rules-of-thumb" that he uses to validate the models' predictions, and gauge the "true health" of the economy, e.g., calling his friend at the Bureau of Labor Statistics to see how many unemployment claims were filed this week in major U.S. cities. He was candid about the limitations of the sophisticated models to predict economic trends, but at the same time defended them, saying, "It is better to be approximately right than certainly wrong."

OR/MS, in general, shares a similar nature with Greenspan's econometric models. Despite the limitations of those models, I think the world is better off with OR/MS practitioners than without us. The world is so full of practical problems, of all sorts, and with businesses having downsized, in some cases to the point of adversely impacting customer service, the urge to "do more with less" beckons the OR/MS solutions that can address such problems.

WHAT IS 'GOOD OR/MS'?

Sometimes, however, I believe we too stringently limit our definition of what passes as "good OR/MS," and prohibit ourselves from taking credit for successes. Loren Platzman, my Probability Models/Queueing Models professor at Georgia Tech, told us graduate students that an OR/MS practitioner's most powerful tools were common sense, good judgment, intuition about a problem's structure, some data points and a spreadsheet. He encouraged us to collect some data, enter it into a spreadsheet, draw some graphs, and "get a feel" for what is happening with the system at hand, to validate your intuition before you build any models. Then, once you have built and validated your models, go and tell someone, and do something about it.

I have followed that rather straightforward advice throughout my career. By the strict definition of what is "good OR/MS," some of my professional work may look, to some people, like "glorified spreadsheets:' However, customers consider them "successes" when the salient model principles, features and functions are implemented, in the form of decision support systems, and ultimately reduce costs or increase revenues by several million dollars.

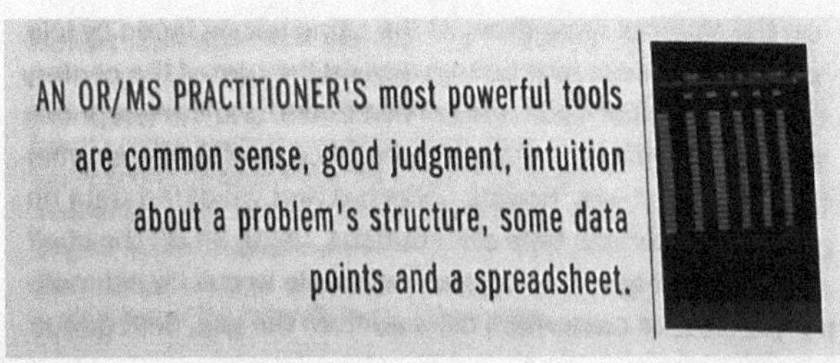

For more than 10 years, I have had the opportunity to apply many different OR/MS methodologies to a lot of different real-world transportation industry problems, ranging from discrete-event simulation analysis of airports to determine capacity of airspace and airfield structures to mixed integer programming heuristics and scheduling algorithms for scheduling aircraft maintenance activities and resources.

I also had the luxury of working in a successful OR/MS organization, so I never really knew what it was like to be in position where doing OR/MS wasn't my primary job function.

OR/MS MINDSET

Recently, I transitioned into a new assignment in the travel distribution technology practice area. I am currently responsible for a group of 60 software engineers focused on building Internet/WWW-based consumer travel reservations systems, and travel agency Internet-enabling software applications. Needless to say, I thought my days as an OR/MS professional were over. However, as I sat in meetings with my team and my new clients wearing my "new IT director hat," something interesting happened. Some very real and perplexing business problems began to arise for which no one in the meetings seemed to possess a solution. However, given my OR/MS training and experience, for me each of the problems fit into a "class of problems" for which model forms were readily available. After I had convinced myself that I was out of the "OR/MS business," I was still able to apply my OR/MS "mindset" and accompanying "bag of tricks," and add value by recommending viable solution approaches to some of the pragmatic business problems at hand. Here are just a few examples:

- My client wanted to better understand the number of people who were being blocked from entering their consumer travel reservations web site, given limited capacity and waiting space to enter the travel transaction processing system. We could simply count the number of "system busy pages" served, but we wanted to know more about the distribution of how many people were arriving at the system. Of those customers who got into the system, how long were they staying? What was the combined effect of arrivals, system capacity and how long people stayed in the system on the number of people blocked from entering? A seemingly complex problem, but one that fits nicely with a classic queueing theory application – the Erlang blocking formula, with Poisson arrivals and exponential service times.

  Electronic commerce and transaction processing systems on the Internet raise many of the same issues faced by telephone engineers (like Erlang) around the turn of the century regarding arrival rates and service times, and the telephone system capacity necessary to balance cost and customer service objectives. Having collected and validated data on arrival and service time distributions, using an off-the-shelf queueing analysis package we were able to quickly estimate the number of customers blocked from the site, or in queue at any given time of the day. This enabled the client to make an informed decision about how many people they would allow into the system to make reservations, and hence generate revenue, versus the cost necessary to provide that system capacity.

- A client wanted to offer a service by which a customer would receive an e-mail message notification in response to an airline fare change reduction of $50 or more on any of their five city-pair airline flight schedule selections, e.g., DFW-LAX. Offering such a service raises several questions regarding IT infrastructure requirements. The number of e-mail messages you will have to send out over a certain time period (e.g., overnight) will drive the need for an e-mail server of a certain capacity. The number of messages sent is based on the number of people who sign up for the service, and how often the fares are likely to change. It turns out that given an estimate of the probability of an airfare change, and some assumptions on the number of people who are likely to sign up for the service, a spreadsheet-based probability model, based on the binomial distribution, can be quickly created to estimate the number of e-mail messages that will need to be sent out (given that a customer will receive a message if one or more their five airfare city-pair selections change).

  This simple model, and subsequent analysis, helped the client determine the cost and benefit parameters associated with providing the service, and enabled them to ensure that the appropriate e-mail server capacity was available to send messages to customers in the timeframe promised.

- One of the "killer apps" in the travel business is to find the "best fare," where "best" depends on customer preferences with respect to price, airline ticket restrictions, time windows and airline choice. The best fare is also limited, of course, to the availability of airline schedule and fare combinations. Given the extremely large number of possible combinations of schedules, airlines and fares, coupled with the fact that the answer must be delivered in real-time (i.e., a travel agent on the phone with a customer, or a consumer on a travel reservations web-site), this turns out to be a very challenging problem for which explicit enumeration is obviously out of the question.

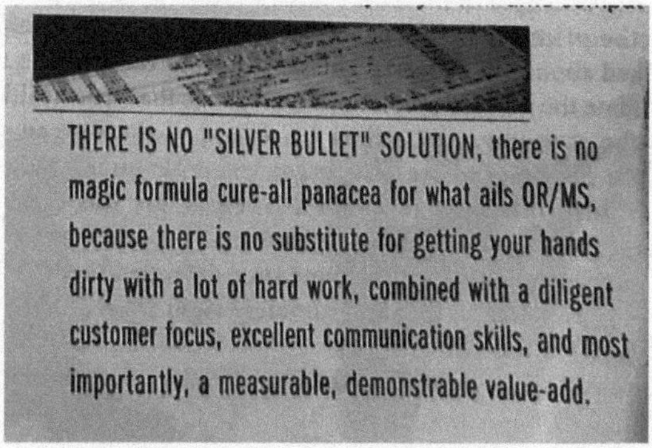
THERE IS NO "SILVER BULLET" SOLUTION, there is no magic formula cure-all panacea for what ails OR/MS, because there is no substitute for getting your hands dirty with a lot of hard work, combined with a diligent customer focus, excellent communication skills, and most importantly, a measurable, demonstrable value-add.

Using branch-and-bound-based implicit enumeration heuristics, combined with a travel agent's knowledge of customer behavior and a lot of heavy-duty mainframe computing power, OR/MS plays a significant role in solving this practical travel industry problem, which is reminiscent of the classic Traveling Salesman Problem, millions of times per day everyday

WHY IS OR/MS IS STRUGGLING?

I discovered in my own career that OR/MS is where you find it. And, more importantly, you don't necessarily have to have an OR/MS job description, or work in an OR/MS department or a consulting company specializing in OR/MS applications to do what I consider "good, valuable OR/MS." The bottom line is that the mathematics to do OR/MS has been there for decades, the computing power is now orders of magnitude better, faster and cheaper than ever before, and certainly the problems and challenges of business abound. So what is missing? Why is OR/MS struggling in a world full of problems that beg to be solved?

My experience tells me that there is no "silver bullet" solution; there is no magic formula cure-all panacea for what ails OR/MS, because there is no substitute for getting your hands dirty with a lot of hard work, combined with a diligent customer focus, excellent communication skills, and most importantly, a measurable, demonstrable value-add. "OR/MS done well" is necessary but not sufficient for OR/MS to be successful. I don't believe that there is anything inherently wrong with OR/MS, other than the way it is applied, or rather, misapplied in many cases. I believe that we, as a community of OR/MS professionals, are struggling with the same challenges as corporations, government, the military, individuals, ad infinitum; that is, in a world of diminishing available resources, what is OR/MS's perceived and demonstrated value-add and return-on-investment? I firmly believe that the customers of OR/MS, whoever and wherever they are, will decide our success or failure for us, based on our performance against that simple metric.

Today, I think of OR/MS less as an academic discipline or a profession in and of itself, and more as a framework for analyzing and solving complex, real-world problems, and making decisions in a structured mathematical way (using a heavy dose of computer and information technology). Like everyone else in the world today, I don't get paid to fulfill a job description, e.g., OR/MS practitioner; I get paid to achieve a bottom line result profitably by solving problems practically as they arise.

BOTTOM LINE BENEFITS

The way to create an effective role for OR/MS in the world depends on champions who are willing to roll up their sleeves, take the time to collect some data, model the problem and explain to management, in English (without Greek letters), the bottom line benefits associated with their solution. Reducing costs, increasing revenues, enhancing customer service, improving efficiency and, most importantly, sticking around long enough to see their ideas get implemented and make a difference is what it's all about. Edison said it first, and any inventor would agree, that "Genius is 1 percent inspiration and 99 percent perspiration." The OR/MS profession is full of inspired geniuses, but we could use a little more perspiration.

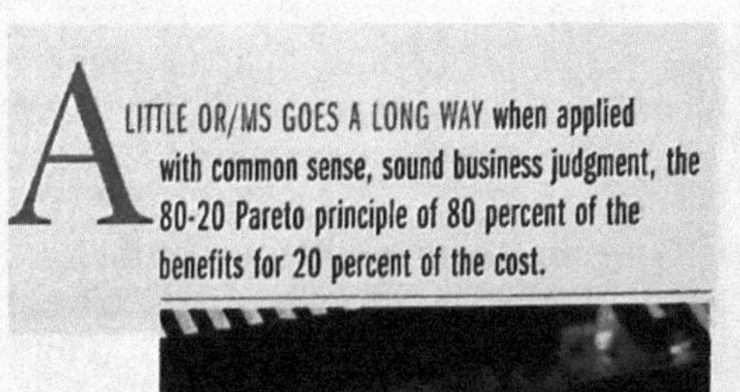

A LITTLE OR/MS GOES A LONG WAY when applied with common sense, sound business judgment, the 80-20 Pareto principle of 80 percent of the benefits for 20 percent of the cost.

The best examples of individuals and companies that apply OR/MS effectively, on a broad scale, are INFORMS Prize winners. If you look at the companies that have won the INFORMS Prize you will find a lot of people who understand as much about business as they do OR/MS. A little OR/MS goes a long way when applied with common sense, sound business judgment, the 80-20 Pareto principle of 80% of the benefits for 20% of the cost. At Sabre Decision Technologies, in particular, we continue to work according to that fundamental principle every day.

All of our senior executives have education in OR/MS, engineering or computer science; more importantly, they all understand the impact of their respective technologies on their respective customers' businesses.

These groups did not spring up over night. There were not suddenly 50, 100, 500 or 1,000 people doing OR/MS projects. They started small as groups of one, three, five or seven analysts and a manager, and grew project by project, deliverable by deliverable, customer by customer, because they added a value so substantial, and so unique that it could not be replaced by the senior management of their respective corporations. These groups have instantiated themselves as a critically necessary and value-added part of their organization's business process.

The best any of us, OR/MS professionals included, can aim for in today's world is to have customers who, at the end of the day, value our contribution enough to hire us for another assignment, and then another after that one. The future of the OR/MS profession will be bright if that is the only criteria for success that ultimately matters.

Douglas A. Gray is a Senior Director of Applications Development and Consulting at SABRE Decision Technologies (SOT), the software and consulting arm of The SABRE Group. Gray has 10 years experience as an OR practitioner, consultant, manager and director at SOT. He has successfully led several decision support application development and consulting initiatives for transportation Industry clients worldwide. Gray currently leads a 60+ member group which is responsible for consumer-direct and travel agency Internet-enabling travel distribution software applications development. He holds a BS degree in Mathematical Sciences from Loyola College, and a master's degree in operations research from the Georgia Institute of Technology's School of Industrial and Systems Engineering. He Is a member of the OR/MS Today Committee.

"The Profession's Paradox: Surrounded by Success," OR/MS Today, June 1997.

In 2016, Dr. Peter C. Bell published a wonderfully insightful article in *OR/MS Today* magazine, entitled "Defining Analytics through the Eyes of Students," that superbly validates the thesis and premise that analytics, and the associated business value and economic impact generated, can be found in a wide range of business problem-solving situations, and does not require super-advanced, sophisticated, mathematical, or technical solutions to be effective, i.e., as he states: *simple, high-level heuristics derived from analytics without doing any mathematics or statistics, and applying these heuristics in a sensible and carefully controlled way that will capture a high percentage of the gains available from using analytics.*

The full article is reprinted in this chapter, but the following insights from Dr. Bell bear repeating:

> Often we appear to be focusing on the development and application of advanced theory or algorithms to try to get a few points closer to the true optimum solution when we know that the data we are using is rough and probably out-of-date, so the result is a really good solution to an approximate problem. My data suggests that sometimes we might do better by focusing on cleaning up the data, improving our understanding of the real issue, and implementing much faster heuristics to find a decent answer to a problem that is closer to reality.
>
> We in analytics tend to think that the 'answer' we derive to an issue is the end-point, but for management it is often a starting point. The 'answer' is usually delivered to a highly intelligent manager (or team) who has a thorough understanding of the business and the issue, and who then merges our analytics work with personal experience and a variety of opinion into planning a path forward. After choosing a course of action, managers implement change, monitor the situation carefully and make corrections. It is common in business strategy to say that the success of a chosen strategy is 'all in the implementation.' The same can be said of analytics; successful analytics can be simple models implemented very effectively.
>
> A quote I use every term to introduce students to analytics comes from Daniel Elwing, former president and CEO of ABB Electric, who said [2]: '[Analytics] is not a project or a set of techniques; it is a process, a way of thinking and managing.'
>
> Along with 'a way of thinking' and a 'process' that starts with data collection, analytics adds value to the data through modeling, which in turns adds value to the decision-maker. Often, very simple models produce substantial benefits.

Dr. Bell is Professor Emeritus, Management Science, at Western University's Ivey School of Business, and taught a course called Competing with Analytics to EMBA students (which is very similar to the course I teach at SMU!). He is a past recipient of the INFORMS Prize for the Teaching of OR/MS Practice.

Although his students' projects rarely involve any analytical heavy lifting, the claimed impact has been impressive with 5% reporting gains in *excess of $5 million*, and 20% reporting gains in *excess of $1 million*!

DEFINING ANALYTICS THROUGH EYES OF STUDENTS

As INFORMS has moved further into "analytics," considerable interest has surfaced in an attempt to define "analytics." Most of the discussion within INFORMS has taken the view that "analytics" has to be highly quantitative, but I hope that INFORMS will take a broader view. Here's why.

Beyond highly quantitative: The case for a broader definition of "analytics." Image © Kheng Guan Toh | 123rf.com

I've been teaching a core analytics course in Ivey's Executive MBA (EMBA) program for many years, and part of this course has always been a project done in the student's workplace where the challenge has been to "do some analytics" that has a positive impact, either financially or organizationally. A key deliverable for this project is a letter from management assessing the impact of the work and its value to the organization. I estimate that I have read and graded almost 1,000 projects over the years.

These are executive student projects (teams of two or three) so they rarely involve any analytical heavy lifting, but the claimed impact has been impressive with about 5 percent of the projects reporting gains in excess of $5 million, and 20 percent reporting gains in excess of $1 million.

Just after I was invited to write an article for this back-to-school issue of OR/MS Today, the following e-mail arrived from a former student:

"I just wanted to follow up on our project from last term. Based on the partial changes we implemented due to the project analytics, we have seen an increase of over $150K, within 30 days of implementation. We took a phased-in approach and it's been successful with negligible complaints or concerns from our users. We are on track to see an additional revenue lift of approximately $1.5M in the next 12 months if all things remain steady.

"Although it may not have met the exact description of an analytics project, we were aware of the risks and the 'stickiness' of our customer base. Demand remains consistent, and is in fact growing with new users."

The second paragraph is a response to my grade report where I expressed doubt as to whether the project met the course requirement to include some "analytics." These students were planning a major change in this company's pricing strategy, but there was no attempt at data collection or "demand modeling," so I expressed the concern that if they started moving prices around, sales might decline so they would need to monitor demand/sales closely.

Many EMBA projects (such as the one above) over the years have applied simple heuristics derived from analytics to real-world situations without doing any math or statistics, and they have claimed a substantial revenue lift or cost reduction. These student projects provide interesting, and I think valuable, lessons about "analytics."

COMPETING WITH ANALYTICS

If I cover a topic in my "Competing with Analytics" course, then in the mind of the student that becomes "analytics." For example, in the classroom we cover several cases where an analytical pricing approach proves useful, and I use these examples to emphasize the basic approach to pricing from analytics, which is to segment the market and then price each segment separately so as to maximize revenue (or contribution) while meeting an overall sales objective or constraint. The models that we build in class also happen to illustrate the potential revenue enhancing value of a basic pricing heuristic "when demand is high price high, and when demand is low price low."

We also discuss a supply chain/pricing case that illustrates the value of pricing decisions in helping out supply chain issues. In this case, the firm greatly benefits from raising the price on products that use bottleneck production processes and reducing the price on products that do not. Again, this leads to a pricing heuristic along the lines that if a product is difficult to schedule, price it high; but if it's easy to schedule, price it low. The cases we cover all include demand data and lend themselves to some statistical analysis and construction of demand models, and can also be set up as optimizations to find optimal revenue maximizing prices. These are executive students, however, and many find the demand modeling and optimization quite challenging.

When students go back to their organizations to do the course project, they remember the general heuristics, and they apply these without doing any data collection, demand modeling and price optimization. I have seen this heuristic pricing approach applied to e-tailing, ready-mixed concrete, long distance transportation, banking, medical services, professional services, graphic arts, manufactured products and many other situations. In all these examples, the claimed revenue gains (supported by "management") were significant and in some cases spectacular.

REASONS FOR PROJECT SUCCESS

There are, of course, many possible explanations for the apparent success of these projects. Perhaps the claimed gains are a mirage? I doubt this to be true in all cases since once these managers have proved that gains are possible, they have often hired analytics people to push these ideas forward. I also receive many e-mails from former students long after the course is over updating me on how "pricing analytics" has transformed their organizations. In some cases I have checked organizational websites and seen market segmentation with variable prices by segment live and in color long after the grades were in.

A second explanation might be that the "before" situation was so ugly that spending a bit of time studying the issue and imposing a slightly less ugly solution produced the observed benefits. If this is the case, then surely this fits one of the claimed benefits of our analytics approach; that analyzing and systematically laying out a problem situation will improve the understanding of a complex issue and will enable the situation to be better managed with the associated benefits.

The explanation that I offer is that sometimes the body of research in analytics can be reduced to a set of high-level heuristics, and that applying these heuristics in a sensible and carefully controlled way will capture a high percentage of the gains available from using analytics.

As an example, if you are selling a service and you identify high- and low-demand market segments (either by time or by customer), and then price the high-demand segments high and the low-demand segments low and jiggle prices around fairly sensibly so as to meet sales targets, you will capture a very high percentage of the potential revenue gains. If you push this further by collecting sales/demand data, building demand models, determining "optimal prices" and installing software to calculate/implement/manage the revenue analytics, you will capture additional revenues, but perhaps not as much as you might expect. If this is the case, the majority of the benefit from the transformation of the firm to an analytics-driven firm comes from the adaptation of an analytics-driven thinking about prices, not from the details of the analytics.

QUICK AND DIRTY

This is not a new idea. Gene Woolsey's book (*Operations Research for Immediate Application: A Quick and Dirty Manual* [1]) advocated and demonstrated that often very simple models applied quickly could produce fast savings and be a hit with management. One difference today is that we have been modeling for some time, and we know that if we model this particular situation the results will generally look like this, so we can implement the result without re-doing the model.

Of course, it's difficult to charge a million dollars for a piece of paper that has on it "price high when demand is high, price low when demand is low," although it might be possible to charge that amount for an extensive data collection/modeling effort that produces this same basic advice. The interesting issue is how much value does the advanced analytics add, over and above a heuristic application of the basic concepts?

Student models that perform optimizations in Excel reinforce the idea that the value of analytics often lies in the approach and not the details. When I look at students' Excel sheets I often find a veritable rats nest of =IF(..), =MAX(..), and /or =VLOOKUP(..) functions inside the optimization model and so the students are trying to optimize potentially highly non-linear problems. Apparently students fiddle around performing multiple runs until they come up with a solution they like, and then they implement the solution leading to a claim of significant benefits. If these students had the advanced analytical skills to properly formulate and optimize their models, I wonder how much the benefits would have increased.

In a similar vein, the project that has claimed the highest benefit (perhaps a one-time saving of $100 million) involved collecting the data necessary to carefully cost out three activities within a major North American company and then optimizing the distribution of workload among these activities. The "advanced analytics" content of this work was a three-variable Excel solver model. The huge benefit of this work clearly came from the analytical approach to issue identification, data collection, careful basic data analysis and costing and the effective implementation of the findings as a new North American "strategy."

These examples strongly suggest that much of the benefit of analytics arises from the analytical problem-solving approach, and while the "advanced analytics" is the cherry on the top, in some (perhaps many) situations, it might be quite a small cherry.

Often we appear to be focusing on the development and application of advanced theory or algorithms to try to get a few points closer to the true optimum solution when we know that the data we are using is rough and probably out-of-date, so the result is a really good solution to an approximate problem. My data suggests that sometimes we might do better by focusing on cleaning up the data, improving our understanding of the real issue, and implementing much faster heuristics to find a decent answer to a problem that is closer to reality.

We in analytics tend to think that the "answer" we derive to an issue is the end-point, but for management it is often a starting point. The "answer" is usually delivered to a highly intelligent manager (or team) who has a thorough understanding of the business and the issue, and who then merges our analytics work with personal experience and a variety of opinion into planning a path forward. After choosing a course of action, managers implement change, monitor the situation carefully and make corrections. It is common in business strategy to say that the success of a chosen strategy is "all in the implementation." The same can be said of analytics; successful analytics can be simple models implemented very effectively.

A quote I use every term to introduce students to analytics comes from Daniel Elwing, former president and CEO of ABB Electric, who said [2]: "[Analytics] is not a project or a set of techniques; it is a process, a way of thinking and managing."

Along with "a way of thinking" and a "process" that starts with data collection, analytics adds value to the data through modeling, which in turns adds value to the decision-maker. Often, very simple models produce substantial benefits.

I encourage INFORMS to seize this broad view of analytics going forward.

REFERENCES

1. G. Woolsey and H. S. Swanson, 1969, 1975, "*Operations Research for Immediate Application: A Quick and Dirty Manual,*" Harper and Row.
2. D.H. Gensch, N.A. Aversa and S.P. Moore, 1990, "A Choice-Modeling Market Information System that Enabled ABB Electric to Expand its Market Share," *Interfaces*, Vol. 20, No. 1 (January/February).

Peter C. Bell
(pbell@ivey.ca)

Peter C. Bell is a professor at the Richard Ivey School of Business at Western University, in London, Ontario, Canada.

Peter C. Bell
Peter C. Bell is a professor at the Richard Ivey School of Business at Western University, in London, Ontario, Canada.

https://pubsonline.informs.org/do/10.1287/orms.2016.04.12/full/

14

O.R. in 2048

Introduction

In 1998, *OR/MS Today* magazine then-editor Peter Horner asked me, and a few other industry professionals, to provide a perspective on what I thought O.R. (operations research) would be 50 years into the future, in 2048. Quite frankly, I ***hate*** making these kinds of predictions because I am not a "futurist," and I have a difficult time imagining what things might be like that far out.

So, at the time of writing this book and collecting the articles, it was late 2023 – 25 years later, or half way through the prediction window. Although there's still 25 more years to go until 2048, my predictions turned out to be directionally correct, so far at least, with some already coming to fruition.

Many of the predictions were actually quite prescient, even now in 2023 as I write this book.

Before you read the 1998 article to see how I and some others did on predicting the future, let's consider some of what has changed since I made my prognostications.

Computing

Computing power – up through 2020 or so – was governed by Moore's law, which stated that CPU speed will double every 18 months and become 50% less expensive over the same period. Moore's law has recently given way to the era of graphical processing units (GPUs) in servers –from companies like NVIDIA, a pioneer in the domain, which enable powerful tools such as OpenAI's large language model (LLM) ChatGPT, trained on 1 trillion parameters. A new AI-based LLM from Amazon called Olympus is being trained on 2 trillion parameters and promises to be even more powerful.

DOI: 10.1201/9781003588344-15

I lived through the major computing eras from mainframes (time-sharing computers) to PCs to engineering workstations to the internet (i.e., network computing) to cloud computing to iLaptops, JUPTYR notebooks, and tablets such as the Apple iPad for UI. But I was nowhere near prescient enough to see GPUs coming!

The availability of cloud-based sandboxes for modeling (versus time-sharing mainframes) has dramatically improved data scientist capacity and capabilities for easily building, testing, and validating powerful models in a fraction of the time historically required. Deploying models as microservices or standalone solutions along with supporting data pipelines has been greatly enabled at high levels of scalability, availability, and reliability using cloud computing. Production deployment of models has created a whole new domain of job functions and roles, including data engineer and MLOps engineer, which did not previously exist.

Models and Algorithms

Since I graduated from undergrad in 1986 and grad school in 1987, there has been a *wave* of new machine learning models invented, such as Gradient Boosting (1999, including XGBoost in 2016) and LASSO Regression (1996), which combine variable selection with model fitting. Tabu search, which was created in 1986 and formalized in 1989, is now regularly and widely used as a heuristic algorithm for solving traveling salesman (TSP) and vehicle routing problems (VRP).

Software Tools

Most impactful to me is the rise in AutoML software tools for data scientists to expedite model fitting and selection. Citizen data scientist desktops, such as DataRobot, Dataiku, H20.ai, and Alteryx, with drag-and-drop GUIs, can remove much of the coding required to handle data, map business process and data flows, and build, test, and validate models.

The rise of Python as a language for data science modeling, challenging industry incumbents such as SAS, SPSS, and even R especially on cost, and providing ample features through rich code libraries and "plug ins," has rapidly evolved and expanded in the past decade or so.

The evolution of cloud computing-based data platforms, from database to data warehouse, to the data lake, data lakehouse, and most recently, the data "mesh," have provided for robust raw and structured data handling and management capabilities in an all-in-one environment.

People

People have evolved quite a bit in the past 25 years, and are more comfortable with using computers at work for problem solving and analysis (albeit mostly in Excel and using SQL queries and dashboards).

Many more people are now comfortable with analytical sciences (albeit mostly in Excel shifting to tools like Python or data science desktops like Dataiku), even those who are not formally trained and educated, but have math, science, or engineering backgrounds and degrees.

The availability of rich, online training capabilities has expanded dramatically, such as those found in Coursera, Udacity, LinkedIn Learning, Udemy, Data Camp, and Kaggle, enabling self-paced learning for those seeking to expand their technical knowledge and strengthen their capabilities.

Unfortunately, most people in business are *still not comfortable* with STEM in general, or math, and AI augmenting for problem solving, decision-making, or question answering, let alone automating their jobs so they can move onto higher order work.

Most people, due to human nature, are *still not any more comfortable with change than they were in 1998.*

My primary thesis about O.R. in 2048 was that the discipline goes "mainstream" and blends in as a "tool set" and "framework" for professionals across the spectrum of business to analyze and solve complex problems. One watershed moment for the commencement of this trend was the publication of *Competing on Analytics* in 2007 (2nd edition, 2017), which heralded analytics as a powerful strategic competitive weapon of business and industry. I predicted that the obscurity of O.R. would persist in 2048, so *long before 2048,* INFORMS shifted their namesake focus to *analytics.* Smart move!

The evolution of O.R. would be fueled by the advancement of data and computing power. Check!

The growth of the field would be ensured by no shortage of complex problems to solve, e.g., omnichannel direct-to-consumer supply chains using optimization software-driven robots and driverless vehicles and drones!

O.R., along with the other analytical sciences, would continue to be enabled, or hindered, by our ability, or inability, respectively, to *market* our solutions and convince executive decision-makers to *invest* in our projects to *realize business value and economic impact, and achieve strategic competitive advantage.*

OR ALMOST GOES MAINSTREAM

Rule #1: The forecast is always wrong.

Rule #2: Only forecast things that will come to fruition or not) after you are either retired or dead.

I strongly believe that OR will still be relevant in 2048, but may not be distinguishable as a separate profession. OR may even be more relevant as computing power increases, enabling the timely solution of larger and more complex problems, and the general population's comfort level with using technology in their daily work increases as well. I believe OR will "blend in" in the real-world as a "tool set" and a "framework" for a variety of professions to leverage in the solution of complex, real-world problems across many industries. I believe that this trend will continue as OR is already being applied in many industries and absorbed into many technology products that I will address later in the article.

While it is extremely doubtful that OR professionals will rule the world, the people who do rule the world (i.e. CEOs, presidents), will have professionals with OR skills advising them in areas such as economics, the military and logistics. Professionals with strong OR skills have risen to high leadership positions in business and government (i.e. Tom Cook at American Airlines, Alan Greenspan at The Federal Reserve Bank). Therefore, I believe OR folks will most definitely continue to influence the world in the future, if not actually rule it.

I strongly believe that technology advances will continue to have a major impact on the OR field in 2048, especially in the areas of computer, information and telecommunications technology. The pervasive use of information technology, including global digital data networks such as the Internet/World Wide Web (WWW), in operating all aspects of both private and public enterprises will drive the need for better tools to enable more effective decision-making and more efficient operations. More information creates an opportunity to make better, more informed decisions, which will drive the need for professionals with strong OR skills to build and deliver such solutions.

The effective application of OR will always rely heavily on computing power and data management technology. The size and complexity of problems that can be solved is a function of Moore's Law, which states that computing power doubles approximately every 18–24 months, while the cost of production continues to fall rapidly. As computers continue to become more powerful, faster and cheaper, OR practitioners will leverage these new technologies to more effectively solve problems that were previously intractable.

Jeremy Rifkin's book, *The End of Work*; postulates the automation of everything, where computer-controlled machinery with built-in, real-time sensing and decision-making algorithms will control everything from steel-making to farming to war-making and many other complex multivariate tasks that humans must do today using manual labor and experience. Several technology advances will greatly influence the pervasiveness of OR methods and techniques being embedded in all sorts of business solutions.OR helps to define how problems should be solved. In 2048 the time frame over which problems are solved will be more critical. As business moves toward real-time transaction processing, motivated by the Internet/WWW, by 2048 real-time decision-making support tools on production lines in factories and distribution channels will be as commonplace as real-time stock trading software is today on Wall Street. The level of pervasiveness will depend more on the value proposition that each technology presents. The more adept OR practitioners are at leveraging advanced computer and information technologies, or working with those knowledgeable in such fields, the more prevalent OR will be.

Advances in laser computing, in which binary bit streams are carried on beams of light, combined with aggressive advances in massively parallel computing, will dramatically increase the size of the problems that can be solved, as well as drastically reduce the time frame over which they are solved.

Global Positioning System (GPS) technology (available today in Hertz rental cars) provides driving directions using an electronic map display, as well as driving routes optimized for the shortest, quickest path between an origin and destination. In the future, such systems will incorporate real-time information on traffic and weather patterns that may influence shortest path calculations, as well as dynamic linehaul and backhaul load information that may alter and optimize route structures "on the fly." Such technology will dramatically improve performance and efficiency in the transportation and distribution industry.

3D computer-visualization and computer-imaging are currently being used in the lumber industry to optimize, in real-time, the cuts of wood carved out of logs moving down a conveyor belt in a sawmill, according to daily market prices. This solution to the cutting stock problem embeds optimization algorithms inside the software that controls the sawblades to maximize profits and minimize waste. I believe that such applications, which combine real-time information and highly advanced computer technology with OR models and algorithms, will be more prevalent as industries are compelled by market forces to become more efficient and more effective through innovative uses of technology.

Optimization models and algorithms are already pervasive in stock portfolio management software, airline management software and supply chain management software. If we extrapolate the trends in just the last 10 years in the manufacturing industry from use of MRP, to MRPII, and now to ERP systems that embed optimization algorithms for solving practical planning, scheduling and control problems (from companies like i2 Technologies, ILOG and Manugistics), the trend is clearly toward a more rigorous, rational, optimization-based approach to factory and distribution system operations management, as opposed to experience-only heuristics.

Search technologies that help customers find exactly what they want on the World Wide Web; online analytical processing (OLAP) tools that help marketers canvass enormous data warehouses to identify trends in customer buying habits; and supply chain management optimization – methodologies that help provide the right product, to the right customer, at the right place, at the right price/time – will be key technologies in 2048 as the world continues to be a competition of the fittest and fastest in meeting the consumer's ever-rising needs!

Those who aren't in the information business will be out of business. If you don't know who your customer is, and you don't have information at your fingertips that tells you all about their needs, wants and desires and about how you can profitably fulfil them better and faster than your competitors, your company will definitely be out of business by 2048, if not well before. Information and how deftly enterprises manipulate information to close the gap between them and their customers, partners and suppliers will be a strategic, competitive advantage.

I believe that we will definitely recognize OR because it will continue to apply many of the same principles that we apply today. The founding principles of OR which provide a rigorous analytical and mathematical framework for solving practical, real-world problems are, in my opinion, so robust that they will remain significant in 2048. OR professionals will continue to do a great job of identifying and solving classes of problems (i.e. traveling salesman, using innovative mathematical models). The limitations of OR will continue to be what they have always been (i.e., computing power, our ability [or inability] to market our "wares" and convincing the decision makers to invest in our solutions to their problems).

I doubt that OR will ever have a name that everyone understands because OR is so broad and pervades so many different business and engineering disciplines (i.e. economics, marketing, industrial engineering). One name cannot possibly capture all of what OR is to so many different people. When outsiders look at OR, they see applications of advanced mathematics (including probability and statistics), combined with advanced information and computer technology, to solve real-world, operational problems. Why should we expect that everyone understand OR in 2048 any more than they understand disciplines like physics and mathematics that have been around for centuries longer than OR? I think OR will become more mainstream as it pervades evolving technologies. However, it is doubtful that OR will ever be recognized as a separate discipline outside the realm of mathematics or systems engineering.

Douglas A. Gray is vice president and chief technology officer of ECWerks Inc., an electronic commerce consulting and software solutions company specializing in supply chain management and distribution technology solutions, based in Tampa, Fla. Formerly, he was an OR practitioner, manager, and director at American Airlines/SABRE Decision Technologies. He has been a member of ORSA (now INFORMS) since 1986.

"OR in 2048: A Flight of Fancy into the Future: OR Almost Goes Mainstream," 1998, OR/MS Today, April.

Conclusion

I will finish the story where I started because the sentiment sums up both my career's intent and legacy.

I have always had a profound appreciation for the power and beauty of mathematics and a penchant and skill for applying a variety of modeling techniques and technologies to solving complex real-world problems in business and industry to deliver tremendous business value and economic impact. My instincts for framing and modeling complex business problems have always been innately strong and continued to develop over time, which I have shared with my teams, colleagues, and students. Developing skills in leadership, communication, people management, project management, and change management took much longer as did designing, building, and deploying large-scale software systems that automate complex business processes.

Intense intellectual curiosity, supported by great intentionality and work ethic, is a cornerstone of a successful career in the analytical sciences. Being driven by a strong desire to fundamentally understand in great detail how things work today in a process or system of any kind and then being tenacious and perseverant in the curiosity of how to make the process or system work *better*, i.e., more efficiently garnering the same or more output (revenue or throughput), with the same or fewer resources (people, vehicles, or other assets), at a lower cost (operating or capital). The most powerful word in your lexicon is *"Why?" Why* does it work that way? *Why* did that outcome occur? Analytical sciences and supporting technology provide you with a wide array of powerful tools to use as economic levers, but your *curiosity* is the fuel that powers those mathematical engines.

I am most proud of the business value and economic impact that my teams have created, which, summing up across the years, is most certainly *more than $3 billion* in cost avoidance and incremental revenue. The awards won by the companies for whom I have worked and the teams I have led are the testaments to this recognizable and significant, measurable, tangible success in creating greater economic efficiency. I believe one should strive to leave things in a better state than they were found and that has been the case with my work in industry.

Like most of us, I started out with nothing but a passion for learning, willingness to work hard, and desire to earn everything based on *merit, trust,* and *strong relationships*. I am a living proof that *anyone* can build a respectable and impactful career based on some fundamental inherent and learned skills and capabilities, concerted and focused effort, and continued growth and development. It is impossible for me to attach too high of a value on my education. My bachelor's and master's degrees in mathematics/statistics from

Loyola University and operations research from Georgia Tech, respectively, are the *foundation* for my entire career in analytical science spanning decades. My MBA from Southern Methodist University (SMU) provided the foundation of understanding how businesses strategize and operate. My "doctorate," however, is from "the school of hard knocks" and is where I learned to successfully lead and manage teams and organizations and deliver large, enterprise-scale system development projects.

It is imperative to understand and always remember that no one, *not one of us*, gets anywhere in the world, in corporate America, academia, or anywhere else, without help from others. My advice would be to not underestimate the value of your "career champions/sponsors" and how important those relationships are to advancement and getting choice assignments. I was "pulled up" the ladder by great leaders, like Mike Parks and Jeff Honeycomb, mentored and made better by others, like Steve Clampett at AA/Sabre and Tom Perkowski at SMU, and lifted up by great architects, engineers, and O.R./data scientists, like Lawrence Hutson (Travelocity), Bobby Johns (AA Maintenance, Travelocity), Rusty Burlingame, Phil Beck, and Mark Song (Southwest Airlines), and Rob Williams (Ontometrics – Blueprint | Optima). Not to mention, or forget, there were countless colleagues, team members, and students along with the way. Any individual should be so very fortunate to have worked for and alongside the caliber of people with whom I have.

Teaching in two notable graduate school programs at SMU reaffirmed in me the belief that you never really understand something until you try to program a computer to do it, or teach others what you *think* you know. Teaching others, at any level, is its own reward. I have learned so many new things like reading books, creating *de novo* curricula, researching and writing cases, lecturing, grading student projects, and interacting with students. There has been no greater satisfaction for me than seeing Executive MBA (EMBA) students who enter a classroom knowing little to nothing about analytics, complete projects, and deliver tangible business value back to their companies in *less than 10 elapsed weeks and four 4-hour class periods! Amazing!*

Although this book has a host of great lessons learned and techniques to be applied, there are two main important chapters on which you should focus moving forward:

- If you are a leader now, or aspire to be one, the most important chapter is **Chapter 11** on **Analytics Leadership Skills**, not so much because the content is earth-shatteringly original, or mind-bendingly sophisticated, but rather, a lot of the skills required are more akin to *running a business* than being a data scientist, which may seem counterintuitive and not so obvious to new leaders. Leading and managing is about *getting other people* to do the right things and do things the right way. Skills as a practitioner are useful to being an effective leader, but they are secondary to all of the other skills listed.

- If you are a practitioner, or aspire to be one, the most important chapter is **Chapter 12** on **Why Data Science Projects Fail**, along with my other co-authored book that expounds on the topic in rich research-based and real-world example-based detail. These are the most common of all of the pitfalls to be aware of and endeavor to avoid in executing projects.

As a practitioner/leader, a few key points of focus include the following:

- *Nothing happens until someone sells something*; true in software sales, where I learned this lesson, but also true for analytics and data science projects; this can be a "pull" or "push" depending on whether a customer is asking for a project (pull) or you are advocating for the project (push); regardless, every project needs a *sponsor* to fund the budget and resources.
- Focus on the *underlying business problem*; understand it as well as, if not better than, your customers and stakeholders before proposing new solution approaches using mathematics.
- Partner closely with your *customers, stakeholders,* and *constituents,* aligning on *every step of the project*, especially the *changes* that will result and their respective *impacts,* always addressing the *why* and *how* with *patience* and *empathy* to ensure a mutual understanding.
- Identify the *metrics and KPIs* that the business cares about in the business problem context and then align your project goals and objectives to favorably impact those measures.
- Many solution approaches to large, complex, real-world problems rarely fit nicely into a single "methodological box."
    - Many solutions comprise *hybrid approaches* that combine *predictive* and *prescriptive* techniques, for example, forecasting demand and then optimizing inventory (e.g., EOQ).
    - Many solutions address system complexity by combining *normative* and *evaluative* approaches, such as mathematical optimization (including heuristics) and Monte Carlo discrete-event simulation.
    - Many solutions are not simply *input-output "black boxes"* but rather interactive, *human-in-the-loop models*, e.g., augmentation AI vs. automation AI.
- Measure the *business value and economic impact* of your model/solution in terms of the metrics and KPIs and convert that to dollars, market share, or NPS improvement or the currency of the problem context.
- Channeling Tom Davenport, *"Models make the enterprise smarter, but models embedded within [high value, mission-critical, planning or real-time] enterprise processes and systems will make the enterprise more economically efficient."* THIS is the *end game*!

Like so many of life's journeys, *your career journey itself is the reward* – the quality and impact of the work you do, the results you achieve and value you deliver, and, most importantly, the people you meet and work alongside. Your career journey in analytical sciences may lead you to one of many potential destinations, depending on what you want to do and what you do best. Trust me, the universe will guide you while you endeavor to make your way. You may want to be an expert individual contributor, e.g., Technical Fellow or Chief Data Scientist; you may want to lead an analytics or technology organization, e.g., Chief Analytics Officer/VP/Director; or you may choose to get an MBA and become a GM, leading a division, large enterprise, or your own company.

Regardless of where you end up in your career, what matters is *not* the destination, office, or title, *not* how much money you make (although I hope you make a lot!) or how many awards you win (or do not win – I am still chasing that elusive Franz Edelman Award!). What matters is that the journey *is* indeed the *reward* as well as the people with whom you travel that road. My advice is to *"become the champion of your own race"*– that is to say, always strive to be better tomorrow than you are today and endeavor to become *indispensable* and endear yourself to those around you. Do not get caught up in endlessly comparing yourself to others –we are all on different paths and move at different paces over time. For a long time, I was greatly bothered by not having has much success as others who were extraordinarily and uniquely successful professionally and financially. If you spend too much time doing this, you will drive yourself a bit crazy, and most likely end up feeling bitter, miserable and less than. Success results from your capability, capacity, will, and, yes, *luck*, e.g., right place, right time, and right solution. Additionally, if one day, you find yourself with more success than Jeff Bezos, Elon Musk, Bill Gates, Larry Ellison, and others, then that will be quite wondrous without a doubt. But if you do not, you can still be a *champion … the champion of your own race …* if you added value, helped your constituents and colleagues, became better every day, and had your share of commensurate professional and financial success along the way.

Hopefully, the lessons presented herein will serve as valuable guideposts on your career journey.

Additionally, while by no means unique to the analytical sciences, I have found the following quotes to be valuable guideposts throughout my career. You may find them useful as well.

> If it can be done, it will be done.
>
> *– Andy Grove, PhD,*
> *former CEO of Intel*

Either by you and your company or your competitors. An appropriate motivation for digital transformation, or economic transformation through analytics, data science, and AI.

> *Audentes Fortuna Iuvat* ("Fortune favors the bold").
>
> – *Latin proverb*

(It is important to note that the bold are not assured fortune, as most of us found out working for ultra-high-risk VC-backed startups, rather, fortune more often falls on those who act boldly.)

You have to take calculated risks in your career, job, project selection, and solution approaches to make a significant impact and break away from the pack. Be bold in digital transformation and economic transformation using analytical sciences.

> *Was mich nicht umbringt, macht mich stärker* (German) *"What does not kill me makes me stronger."*
>
> *–Part of aphorism number 8 from the "Maxims and Arrows" section of Friedrich Nietzsche's* ***Twilight of the Idols*** *(1888).*

I strongly identify with the fundamental principle of *existentialism*, which promotes the belief that *existence precedes essence*. That who you are and that which you are becoming is not *predetermined* but is a function of your experiences gained throughout your existence. You will be challenged, you will struggle, you will fail, but as long as you keep getting up whenever you are knocked down, and *your defeat doesn't kill you*, then you will be stronger for the next fight.

> Luck is the residue of design.
>
> – *Originally attributed to English poet John Milton, the quote is culturally attributed to Branch Rickey, GM of the Brooklyn Dodgers who brought Jackie Robinson to Major League Baseball.*

Good things happen to those who plan, fail to plan, and plan to fail. Then, as a result of planning and preparedness, the ball tends to bounce your way more often.

> *Sic transit gloria mundi* (Latin) ("Thus passes worldly glory") or All glory is fleeting.
>
> – *The latter phrase, often attributed to Napoleon, was meant to imply that fame or glory is transient, but when someone is forgotten it is forever.*

People will remember your greatest accomplishments for a short time, so you need a string of successes to propel your career forward and increase your trajectory. (The old saws "What have you done for me lately?" and "Don't rest on your laurels" come to mind!)

Famously, the last line of the multiple Academy Award-winning 1970 movie *Patton* (one of my favorite movies of all time, certainly in the military history genre) about arguably the greatest military field commander of the 20th century whose controversial statements and actions led to his diminishment from WWII history despite his extraordinary battlefield leadership, exploits, and successes leading corps and armies in North Africa and Europe to defeat Nazi Germany.

The "entropy of the universe and the effect of gravity" causes one's fortunes to rise and wane – you will not win every time, but that should not ever stop you from leaning in to new challenges.

References and Bibliography

Original Articles

1. From *OR/MS Today*, October 1991 Issue, *Graphic Evidence*. "The Dual Challenge of the OR Practitioner."
2. From *OR/MS Today*, December 1992 Issue, *Transportation*. "Airworthy."
3. From *OR/MS Today*, December 1993 Issue, *Simulation: A Survey of Flexible Tools for Modeling*. "Broaden Perspective: Consulting concepts come to life for author while working on American Airlines maintenance project."
4. From *OR/MS Today*, April 1994 Issue, *Strategic Supply Chain Management*. "Right Tool, Place, Time."
5. From *OR/MS Today*, October 1994 Issue, *The Future of Manufacturing*. "Under Fire: Lessons from the Front."
6. From *OR/MS Today*, June 1997 Issue, *The Profession's Paradox*. "Surrounded by Success."
7. From *OR/MS Today*, April 1998 Issue, *OR in 2048: A Flight of Fancy into the Future*. "OR Almost Goes Mainstream."
8. From *Analytics* magazine (INFORMS), a 12-part series (2023–2024), entitled "Why Data Science Projects Fail: Practical Principles for Data Science Success."

Article Posts on LinkedIn

1. Advanced Analytics Commodity Case
2. Advanced Analytics is Economically Transformational
3. Digital Transformation
4. Analytics Soft Skills
5. Top 10 Analytics Leadership Skills

Index

Note: *Italic* page numbers refer to figures.

FSC
www.fsc.org
MIX
Papier | Fördert gute Waldnutzung
FSC® C083411

Kolibri 360 GmbH
Ferdinand-Jühlke-Straße 7
99095 Erfurt, Deutschland
produktsicherheit@kolibri360.de